SERIES IN COMMUNICATION AND CONTROL SYSTEMS

Series in Communication and Control Systems

A. V. Balakrishnan, A. A. Dorodnitsyn, & J.-L. Lions, eds.
Vistas in Applied Mathematics: Numerical Analysis, Atmospheric Sciences, Immunology
1986, 396 pp., ISBN 0-911575-38-3

R. E. Kalman, G. I. Marchuk, A. E. Ruberti, & A. J. Viterbi, eds.
Recent Advances in Communication and Control Theory
1987, 500 pp., ISBN 0-911575-46-4

A. V. Balakrishnan
Kalman Filtering Theory
1988, 267 pp., ISBN 0-911575-49-9
2nd, rev. & enl., ed.

Translations Series in Mathematics and Engineering

M. I. Yadrenko
Spectral Theory of Random Fields
1983, 267 pp., ISBN 0-911575-00-6

G. I. Marchuk
Mathematical Models in Immunology
1983, 378 pp., ISBN 0-911575-01-4

A. A. Borovkov, ed.
Advances in Probability Theory: Limit Theorems and Related Problems
1984, 392 pp., ISBN 0-911575-03-0

V. A. Dubovitskij
The Ulam Problem of Optimal Motion of Line Segments
1985, 128 pp., ISBN 0-911575-04-9

N. V. Krylov, R. Sh. Liptser, and A. A. Novikov, eds.
Statistics and Control of Stochastic Processes
1985, 521 pp., ISBN 0-911575-18-9

Yu. G. Evtushenko
Numerical Optimization Techniques
1985, 575 pp., ISBN 0-911575-07-3

V. F. Dem'yanov and L. V. Vasil'ev
Nondifferentiable Optimization
1985, 472 pp., ISBN 0-911575-09-X

A. A. Borovkov, ed.
Advances in Probability Theory: Limit Theorems for Sums of Random Variables
1985, 313 pp., ISBN 0-911575-17-0

V. F. Kolchin
Random Mappings
1986, 224 pp., ISBN 0-911575-16-2

L. Telksnys, ed.
Detection of Changes in Random Processes
1986, 240 pp., ISBN 0-911575-20-0

V. F. Dem'yanov and A. M. Rubinov
Quasidifferential Calculus
1986, 301 pp., ISBN 0-911575-35-9

V. P. Chistyakov, B. A. Sevast'yanov, and V. K. Zakharov
Probability Theory for Engineers
1987, 175 pp., ISBN 0-911575-13-8

B. T. Polyak
Introduction to Optimization
1987, 464 pp., ISBN 0-911575-14-6

V. V. Ivanishchev and A. D. Krasnoshchekov
Control of Variable Structure Networks
1987, 154 pp., ISBN 0-911575-05-7

R. F. Gabasov, and F. M. Kirillova
Methods of Optimization
1988, 394 pp., ISBN 0-911575-02-2

Comcon Conferences Proceedings

Advances in Communication & Control Systems. Proceedings of the First International Conference, Washington, D.C., June 1987
Edited by N. DeClaris
1988, 224 pp., ISBN 0-911575-47-2

A. V. BALAKRISHNAN

STATE SPACE THEORY OF SYSTEMS

AN INTRODUCTION

2nd Edition
Revised and Enlarged

OPTIMIZATION SOFTWARE, INC.
PUBLICATIONS DIVISION, NEW YORK

Author
A.V. Balakrishnan
Electrical Engineering Department
School of Engineering and Applied Science
University of California
Los Angeles, California

Library of Congress Cataloging in Publication Data

Balakrishnan, A.V.
State space theory of systems.

(Series in communication and control systems)
Rev. ed. of: Elements of state space theory of systems. c1983.
Includes bibliographies and index.
I. System analysis. 2. State-space methods.
I. Balakrishnan, A.V. Elements of state space theory of systems. II. Title. III. Series.
QA402.B33 1988 003 88-25591
ISBN 0-911575-54-5

Printed in the United States of America

First edition 1983: *Elements of State Space Theory of Systems*

About the Author:

A.V. Balakrishnan is Professor of Electrical Engineering and Professor of Mathematics at UCLA, and Director of the Flight Systems Research Laboratory. He is a Fellow of the IEEE, and past Chairman of the IEEE Information Theory Group. He received a Certificate of Recognition from the National Aeronautics and Space Administration for his scientific contributions to processing of Flight Test data. He was awarded the Guillemin Prize for his many contributions to Communication and Control Systems Engineering. He is also the recipient of the Silver Core Award from the International Federation of Information Processing, and more recently, of a Group Achievement Award from NASA Langley FRC for his leadership role in the Spacecraft Control Laboratory Experiment (SCOLE). He is the author of several books and monographs in Applied Mathematics and Engineering, and serves on the Editorial Boards of numerous journals and monograph series. Currently he serves as President of the Comcon Board.

TABLE OF CONTENTS

PREFACE TO THE FIRST EDITION

This text is intended for a First Course in State Space Theory of Linear Systems for first-year graduate students. The emphasis is on basic theory and concepts rather than practical applications or computer simulation. A course on Linear Algebra is a firm prerequisite. The first (zcroth) chapter merely provides a review of relevant essential material on matrix theory used in the sequel.

The notions of Input-Output and State are introduced in Chapter 1. Chapters 2 and 3 are devoted to Controllability and Observability, respectively, and Chapter 4 to Stability. After a brief look at canonical representation in Chapter 5, we go to State Estimation and State Stabilizability by Feedback in Chapter 6. The discussion throughout is limited to continuous-time linear time-invariant dynamic (finite-dimensional) systems. Because of its importance in digital computer implementation, the discrete-time case is treated in Chapter 8.

To fit within a one-quarter to one-semester limit, much material has had to be eschewed, even at the risk of making the selection appear arbitrary. The objective was to specify minimal content: a syllabus essential for later courses.

Contrary to current trend, no attempt at "spoon-feeding" has been made. The reader may have to supply appropriate "intermediate steps" in arguments and thereby, hopefully, increase his or her understanding and augment his or her skills.

In addition to the worked-out examples, a small list of problems is appended to each chapter; and these should be regarded as an integral part of the chapter.

PREFACE TO THE SECOND EDITION

In preparing the second edition I have taken advantage of the opportunity to make changes, drawing in particular on the experience gained in class-room use of the first edition. The material has been revised throughout and a new chapter added.

The first chapter—the review chapter on matrix theory—has been expanded to make it more self-contained and more responsive to the needs of later chapters. The treatment is deliberately kept elementary: thus matrices are assumed to be simple where necessary to shorten proofs. Contour integrals are no longer invoked, so that little or no knowledge of complex variable theory background is required. While essential notions from linear space theory—such as range space and null space, are introduced, abstraction is kept to a minimum. It is assumed that the student is conversant with the basics of matrix algebra (for example, calculating determinants by elementary row-column operations) and, similarly, with linear differential equations, as in a customary course in linear systems.

We begin in Chapter 2 with the system as a black box: a collection of input-output pairs, and introduce the state as a secondary or derived notion, which at first level removes the multivaluedness of the output. The bulk of the chapter is devoted to deducing states from input-output relations and vice versa. The multiplicity of possible state spaces prepares the way for the concepts to follow in succeeding chapters. The first of these is controllability, the subject of Chapter 3. It is introduced as a state space notion, and the equivalence to the classical convolution integral definition of a linear system is stressed. Observability is introduced in Chapter 3 in the context of estimating states from input-output pairs. The state reduction process is treated as well as the isomorphism between observable state spaces. The third of the triad of concepts, stability, is treated in Chapter 4. It includes in particular a derivation of the exponential bound based on subadditivity.

More advanced topics are taken up in Chapter 6, such as canonical forms for the controllable part of the state space, controllability and pole-zero cancellation, as well as arbitrary pole-placement by appropriate control, specialized to single input-single output systems. As in any introductory volume, choices have to be made in introducing applications. The bias here is toward filtering and control theory, rather than circuit analysis. Thus Chapter 7 treats state estimation and stabilization by feedback control showing the role played by, and the necessity of, the state-space notions studied in the previous chapters. Included in Chapter 7 is a class of robust controllers in the framework of vibrating system models in mechanical structures. In particular, this should underscore the interdisciplinary nature of state-space theory. The treatment is extended to linear time-varying systems in Chapter 8, where appropriate generalization of the notions of controllability and observability is developed. Because of their importance in digital control systems, as well as in simulation, a separate chapter, Chapter 9, is devoted to discrete-time systems. Finally, Chapter 10 examines state-space theory for systems which are not linear, lest the student think that all systems are linear! The mathematical level naturally escalates here, but the treatment is again kept elementary and engineering oriented. Examples are drawn from engineering mechanics, with emphasis on stability and stabilizability by feedback control.

Each chapter concludes with a set of problems, in addition to those scattered throughout the body of the chapter to provide immediate illustration of specific concepts introduced. They still leave much room for the instructor to supplement them with problems closer to engineering practice. The intent here is, rather, for the student to understand the concepts and acquire related analytical techniques for future use.

The overall presentation follows a systematic approach with logically sequenced topics in contrast to a purely problem-solving "bag of tricks" approach. In this sense it is more mathematical.

It is a pleasure to acknowledge the help of many students and colleagues in the course of this work, and in particular Professor N. Levan and N. Do for their assistance in removing errors and improving readability.

CHAPTER 1
REVIEW OF MATRIX ALGEBRA

We begin with a brief review of miscellaneous results in matrix algebra that we shall need in the sequel. The reader should consult the references listed for more details; especially since not all assertions recalled are proved.

Preliminary Notions and Definitions

Let $\mathcal{L}$ denote the class of all $m \times n$ matrices. When we need to call attention to the components (equivalently, "entries" or "terms") of a matrix we use the notation:

$$A = \{a_{ij}\}, \qquad 1 \le i \le m; \ 1 \le j \le n .$$

If we fix j, and let i run, we have a row, the jth row. If we fix i and let j run, we have a column, the ith column. An $m \times 1$ matrix is referred to as a "column" vector or simply a vector; similarly, a $1 \times n$ matrix is called a "row" vector, or simply a vector. Although we shall allow the components $\{a_{ij}\}$ to be complex-valued, we shall for the most part deal only with real-valued or real matrices—all of whose components are real. Following convention we shall use the term "scalars" to denote numbers, real or complex, distinguishing them from "vectors." An $m \times n$ matrix is a "rectangular" matrix; when $m = n$, we use the term "square" matrix to indicate $m = n$.

The adjoint of a matrix A, denoted A^*, is defined by

$$A^* = \{\bar{a}_{ji}\}, \qquad 1 \le j \le n; \ 1 \le i \le m ,$$

where

$$A = \{a_{ij}\}, \qquad 1 \le i \le m; \ 1 \le j \le n ,$$

and the superbar denotes complex conjugate. Note that A^* is $n \times m$. It is customary to define the "transpose" of A, denoted A^t, by

$$A^t = \{a_{ji}\} , \qquad 1 \le j \le n; \; 1 \le i \le m .$$

We shall only use the adjoint, and of course $A^t = A^*$ for real matrices.

If A, B are two matrices in $\mathcal{L}$:

$$A = \{a_{ij}\} ; \qquad B = \{b_{ij}\} ,$$

and α, β are scalars, by

$$\alpha A + \beta B$$

we mean the matrix with components

$$\{\alpha a_{ij} + \beta b_{ij}\} .$$

Thus defined $\mathcal{L}$ is a "linear vector space." Any subset (of $\mathcal{L}$) which is also a linear vector space is called a "linear subspace." The set consisting of the zero element is a linear subspace. Recall the definition of matrix multiplication: If A is $m \times n$ and B is $n \times p$, then we define

$$C = AB$$

as the $m \times p$ matrix with coefficients

$$C = \{c_{ij}\} ; \qquad c_{ij} = \sum_{k=1}^{n} a_{ik} b_{kj} .$$

While multiplication is not commutative (even for square matrices), it is associative:

$$A(BC) = (AB)C .$$

Also

$$(AB)^* = B^* A^* .$$

Let A be a square matrix, say $m \times m$. By the trace of a matrix (defined only for square matrices) we mean the sum of the "diagonal" terms:

$$\text{Tr } A = \sum_{1}^{m} a_{ii} .$$

We refer to a_{ij}, $i \neq j$ as the "off-diagonal" terms. By a "diagonal" matrix we mean a matrix in which the off-diagonal entries are zero. For any rectangular matrix A, we note that both AA^* and A^*A are square matrices. It is easy to see that

$$\operatorname{Tr} AA^* = \operatorname{Tr} A^*A = \sum_{i=1}^{m} \sum_{j=1}^{n} |a_{ij}|^2 .$$

We define the norm of a matrix (or "matrix" norm) by

$$\|A\| = \sqrt{\operatorname{Tr} AA^*} = \sqrt{\operatorname{Tr} A^*A} .$$

Note that if $\|A\| = 0$, A must be the zero matrix—with all entries zero.

An important concept we shall need is that of "inner product" of matrices. (The older and now obsolete term is "dot-product.") For any two matrices A, B in $\mathcal{L}$ we define

$$[A, B] = \operatorname{Tr} AB^* ,$$

the right side being well-defined since AB^* is square ($m \times m$). Note that

$$[B, A] = \overline{[A, B]} ,$$

$$[A, A] = \|A\|^2 \geq 0 ,$$

$$[\lambda A, B] = \lambda[A, B] .$$

An oft-needed inequality for us is the Schwarz inequality for inner products:

$$[A, B] \leq \|A\| \, \|B\| .$$

It can be proved as follows: for any complex number λ, and A, B in $\mathcal{L}$,

$$0 \leq \|A + \lambda B\|^2 = [A, A] + \lambda[B, A] + \overline{\lambda}[A, B] + [B, B]|\lambda|^2 .$$

If $\|B\| = 0$, the inequality is trivially true. Hence let

$$\lambda = -\frac{[A, B]}{\|B\|^2} , \qquad \|B\| \neq 0 .$$

Substituting this value of λ, we have

$$0 \leq [A, A] - \frac{|[A, B]|}{\|B\|^2} .$$

Or

$$|[A, B]|^2 \leq \|A\|^2 \, \|B\|^2$$

yielding the Schwarz inequality upon taking the square root on both sides.

Note in particular that equality holds if and only if

$$A = \gamma B$$

for some scalar γ. From the Schwarz inequality we have that

$$\begin{aligned} \|A+B\|^2 &= \|A\|^2 + \|B\|^2 + [A, B] + [B, A] \\ &\le \|A\|^2 + \|B\|^2 + 2\|A\|\,\|B\| \\ &= (\|A\| + \|B\|)^2 \end{aligned}$$

and hence

$$\|A+B\| \le \|A\| + \|B\| ,$$

known as the triangle inequality.

A useful notion deriving from that of "inner-product" is that of "orthogonality." Thus we say that A is orthogonal to B (or sometimes also "perpendicular" to B because of the analogy with the concept in Euclidean geometry) if

$$[A, B] = 0 .$$

They are said to be "orthonormal" if they are orthogonal and have unit norms:

$$[A, B] = 0 ; \qquad \|A\| = \|B\| = 1 .$$

Let us specialize to the class of $n \times 1$ vectors. We denote it R^n when the scalars are real-valued and C^n when we want to emphasize that the scalars are complex-valued.

The vectors ("coordinate basis" vectors):

$$e_i = \{a_{ij}\} , \qquad i = 1, \ldots, n ,$$

where

$$\begin{aligned} a_{ij} &= 1 , \qquad i = j , \\ &= 0 , \qquad i \ne j , \end{aligned}$$

are clearly orthonormal. Moreover, we can express any x in C^n by

$$x = \sum_1^n [x, e_i] e_i .$$

This representation holds for any n orthonormal vectors—they form a "basis." In particular

$$[x, y] = \sum_1^n [x, e_i] [y, e_i] ,$$

and

$$\|x\|^2 = \sum_1^n |[x, e_i]|^2 .$$

Let S denote a subset of C^n. By the orthogonal complement of S, denoted $S^{\perp}$, we mean the set consisting of all vectors x such that x is orthogonal to every vector in S. Note that $S^{\perp}$ is a linear space; in particular $C_n^{\perp}$ consists of the zero vector only. If S, T are two sets in C^n, by $S + T$ we mean the set of all elements of the form $x+y$, where $x \in S$, $y \in T$. Note that if $\mathcal{A}$ is a linear subspace, then

$$C^n = \mathcal{A} + \mathcal{A}^{\perp} .$$

So far we have basically looked on an $m \times n$ matrix as an $mn \times 1$ vector. We may also look upon them as linear transformations or operators. Thus for A in $\mathcal{L}$ and any $n \times 1$ column vector x, we can define the function of x by:

$$f(x) = Ax$$

by matrix multiplication, mapping the space C^n into C^n. That is a *linear* transformation (or operator) in the sense that

$$A(\alpha x + \beta y) = \alpha Ax + \beta Ay$$

for any scalars, α, β and $n \times 1$ vectors x, y.

We define the "operator norm" by

$$\|A\|_o = \max_{\|x\| \leq 1} \|Ax\| .$$

Thus defined we can readily see that for any scalar λ

$$\|\lambda A\|_o = |\lambda| \, \|A_o\|$$

and that

$$\|A + B\|_o \leq \|A\|_o + \|B\|_o \qquad \text{(triangle inequality)} ,$$

which follows from

$$\|(A + B)x\| = \|Ax + Bx\| \leq \|Ax\| + \|Bx\| .$$

Since for any nonzero column vector x we have

$$\left\| \frac{x}{\|x\|} \right\| = 1 ,$$

we can write:

$$\|Ax\| = \frac{\|Ax\|}{\| x \|}\|x\| = \left\| A\left(\frac{x}{\|x\|}\right) \right\| \|x\|,$$

and since

$$\left\| A\left(\frac{x}{\|x\|}\right) \right\| \leq \|A\|_o ,$$

we have:

$$\|Ax\| \leq \|A\|_o \|x\| .$$

Connecting inner products and adjoints, note that

$$[Ax, By] = [B^* Ax, y] = [x, A^* By] .$$

Let $\{e_i\}$, $i = 1, ..., n$ be an orthonormal basis in the space of $n\times 1$ vectors. Then for each i, $e_i e_i^*$ is an $n\times n$ matrix, and it is not difficult to verify that

$$\sum_1^n e_i e_i^* = \text{Identity Matrix}^\dagger ,$$

using the fact that for every $n\times 1$ vector x

$$\sum_1^n e_i e_i^* x = \sum_1^n [x, e_i] e_i = x .$$

In particular therefore for any $n\times n$ matrix A we have

$$\text{Tr}\, A = \sum_1^n [Ae_i, e_i]$$

since

$$\sum_1^n [Ae_i, e_i] = \text{Tr}\, A \sum_1^n e_i e_i^* = \text{Tr}\, A .$$

† Denoted by I, regardless of dimension, throughout.

Further, for any A in $\mathcal{L}$ and any $n \times 1$ vector x, we have

$$\begin{aligned} \|Ax\|^2 &= [Ax, Ax] = [A^*Ax, x] \\ &\leq \|A^*Ax\| \, \|x\| \qquad \text{(Schwarz Inequality)} \\ &\leq \|A^*A\|_o \, \|x\|^2 . \end{aligned}$$

Hence

$$\|A\|_o \leq \sqrt{\|A^*A\|_o} \ .$$

On the other hand we have:

$$\begin{aligned} \|Ax\|^2 &= \sum_1^n |[Ax, e_i]|^2 = \sum_1^n |[x, A^* e_i]|^2 \\ &\leq \sum_1^n \|A^* e_i\|^2 \, \|x\|^2 \qquad \text{(Schwarz Inequality)} \\ &= \sum_1^n [A^* e_i, A^* e_i] \, \|x\|^2 \\ &= \|x\|^2 \sum_1^n [AA^* e_i, e_i] \\ &= \|x\|^2 \operatorname{Tr} AA^* . \end{aligned}$$

Hence

$$\left\| A \frac{x}{\|x\|} \right\|^2 \leq \operatorname{Tr} AA^*$$

or

$$\|A\|_o \leq \sqrt{\operatorname{Tr} AA^*} = \|A\| .$$

The operator norm is thus less than or equal to the (matrix) norm. For a diagonal matrix

$$\|A\|_o = \max_i |a_{ii}| \leq \|A\| = \sqrt{\sum_1^n a_{ii}^2} \ ,$$

showing that the difference can be large.

Problems

1. Let A, B be square matrices. Show that

$$\text{Tr } AB = \text{Tr } BA .$$

2. Prove the associative property for matrix multiplication:

$$ABC = A(BC) = (AB)C ,$$

where

$$A \text{ is } m \times n ,$$

$$B \text{ is } n \times p ,$$

$$C \text{ is } p \times q .$$

Show that

$$(ABC)^* = C^* B^* A^* .$$

3. Produce a counterexample to show that matrix multiplication is not commutative, even for square matrices.

4. For any set S in C^n, show that $(S^\perp)^\perp$ is a linear vector space containing S. Show that

$$C^n = S^\perp + (S^\perp)^\perp .$$

What is

$$((S^\perp)^\perp)^\perp \ ?$$

Square Matrices: Spectral Theory

Let us now specialize to the case of square matices—say $n \times n$. We shall denote this class by $\mathcal{L}$.

For the purposes of Spectral Theory it is convenient to consider $\mathcal{L}$ as the space of linear operators on the space C^n of $n \times 1$ vectors with the components allowed to be complex, the letter C denoting this, mapping C^n into C^n.

We need to recall some basic features of the space C^n. First of all, C^n is an "n-dimensional linear vector space." Given a finite number of vectors v_i, $i = 1,...,m$ in C^n, we say they are linearly dependent if we can find scalars $\{\alpha_i\}$, not all zero, such that

$$\sum_1^m \alpha_i v_i = 0 .$$

Otherwise they are said to be linearly independent. The coordinate vectors

$$e_i = \{a_{ij}\} , \qquad i = 1, ..., n ,$$

where

$$a_{ii} = 1 ,$$

$$a_{ij} = 0 , \qquad i \neq j ,$$

being orthogonal, are readily seen to be linearly independent. A convenient test for dependence is to invoke the determinant. Recall the definition of the determinant of a square matrix. We shall denote the determinant of A by Det.A.

Given a set of n vectors in C^n, v_i, where

$$v_i = \{a_{ij}\} , \qquad i = 1,...,n ; \ j = 1,...,n ,$$

the right side defines a square matrix. The vectors $[v_i]$ are linearly dependent if and only if the determinant of this matrix is zero. More generally, given any m *vectors* v_i, i, $= 1,...,m$, we may look at the determinant of the "moment" matrix defined by

$$\{[v_i, v_j]\} , \qquad i = 1,...,m ; \ j = 1,...,m .$$

Linear dependence is equivalent to the determinant vanishing. We may also note that any set of $(n+1)$ (or more) vectors in C^n must be linearly dependent.

Given any m vectors $\{v_i\}$ in C^n, the class of vectors of the form

$$\sum_1^m a_k v_k ,$$

where the $\{\alpha_k\}$ are arbitrary scalars—"linear combinations of the vectors v_i"—is a linear vector space by itself; it is referred to as a "subspace" of C^n "spanned" by the vectors v_i, $i = 1,...,m$.

For any A in $\mathcal{L}$, a nonzero vector x in C^n such that for some scalar λ:

$$Ax = \lambda x \tag{1.1}$$

is called an "eigenvector" corresponding to the "eigenvalue," λ. Since if x

is an eigenvector, so clearly also is γx for any nonzero scalar γ, we usually normalize an eigenvector to have unit norm. It is important to note that eigenvectors corresponding to distinct eigenvalues are linearly independent. In fact, if

$$Ax_i = \lambda_i x_i , \qquad i = 1,\ldots,m , \; x_i \neq 0 ,$$

and

$$\sum_1^m \alpha_k x_k = 0 ,$$

we can write (since not all α_k are zero)

$$x_1 = \sum_2^m \beta_k x_k$$

and

$$Ax_1 = \sum_1^m \beta_k A x_k = \sum_2^m \beta_k \lambda_k x_k = \lambda_1 x_1 = \sum_2^m \beta_k \lambda_1 x_k ,$$

and hence, subtracting, we have

$$\sum_2^m \beta_k (\lambda_k - \lambda_1) x_k = 0 .$$

Not all β_k can be zero, since in that case $x_1 = 0$; and since the eigenvalues are distinct, it follows that the eigenvectors x_2, ..., x_m are linearly dependent. Proceeding in this way we would reduce to

$$\beta_m (\lambda_m - \lambda_{m-1}) x_m = 0 ,$$

and since $x_m \neq 0$, we would arrive at a contradiction, thus proving the linear independence.

Next let γ be an eigenvalue of the $n \times n$ matrix A with eigenvector x:

$$Ax = \gamma x , \qquad x \neq 0 ,$$

which we can rewrite as

$$(\gamma I - A)x = 0 , \qquad x \neq 0 .$$

Let v_i, $i = 1,\ldots,n$ denote the column vectors of $(\gamma I - A)$ and x_i, $i = 1,\ldots,n$ denote the components of x.

Then

$$(\gamma I - A)x = \sum_1^n x_i v_i = 0 , \qquad \text{not all } x_i = 0 .$$

Hence one of the columns v_i can be expressed as a linear combination of the others and hence

$$\det (\gamma I - A) = 0 .$$

Conversely, by reversing the argument, this condition will imply in turn that γ must be an eigenvalue of A.

It is easy to see that the eigenvalues are precisely the roots of the "characteristic polynomial" $\Delta(\lambda)$

$$\Delta(\lambda) = \text{Det.}(\lambda I - A) = 0 .$$

We can verify that $\Delta(\lambda)$ is a polynomial of degree n in λ, the coefficient of x being unity.

A matrix is said to be "simple" if its eigenvectors span the space C^n. Note that a sufficient condition for a matrix to be simple is that all its eigenvalues are distinct. One can make the argument resting on "computational" grounds that eigenvalues of any matrix must be distinct; they are equal only because of computational error. In any event, many "proofs" become simpler if we consider only simple matrices.

Let A be simple and $\{e_i\}$ denote its eigenvectors and λ_i the corresponding eigenvalues. Let us take $\{e_i\}$ as column vectors. Then we have that

$$AE = ED ,$$

where

$$E = |e_i, \ldots, e_n| ,$$

$$D = \text{diag. } \{\lambda_i\} ;$$

and since E must be nonsingular, we have that

$$E^{-1} AE = D ,$$

or A can be "diagonalized." Thus "diagonalizability" holds for simple matrices, although this is not true if the matrix is not simple, in general.

Let $\{\lambda_i\}$ denote the eigenvalues of a matrix A. Then we define

$$\max_i |\lambda_i| = r = \text{"spectral radius"} .$$

Clearly

$$r \leq \|A\|_o .$$

More importantly (see below for proof),

$$\log r = \lim_{n \to \infty} \frac{\log \|A^n\|_o}{n} . \tag{1.2}$$

It is quite possible that

$$\|A\|_o > r .$$

For example, if

$$A = \begin{vmatrix} 0 & 1 \\ \frac{1}{4} & 0 \end{vmatrix} ,$$

we have

$$r = \frac{1}{2} ,$$

while

$$\|A\|_o = 1 ,$$

as is evident by taking

$$\frac{\|Ax\|}{\|x\|} \quad \text{for } x = \begin{vmatrix} 0 \\ 1 \end{vmatrix} .$$

A matrix A is "self-adjoint" if

$$A = A^*$$

(if real, "symmetric").

The eigenvectors of a self-adjoint matrix are orthogonal. The subspace of eigenvectors corresponding to the zero eigenvalue (the null space) is orthogonal to the range space. In particular it follows that any self-adjoint matrix is simple; in fact we have the "spectral representation"

$$Ax = \sum_1^n [x, e_i]\lambda_i e_i ,$$

where $\{e_i\}$ are orthonormalized eigenvectors of A, which form a basis for R^n.

From

$$\|Ax\|^2 = [Ax, Ax] = \sum_1^n \lambda_i^2 |[x, e_i]|^2$$
$$\leq \max_i \lambda_i^2 \sum_1^n |[x, e_i]|^2$$

it follows that

$$\|A\|_o = \text{Spectral Radius of } A .$$

A matrix A is nonnegative definite if

$$[Ax, x] \geq 0 \qquad \text{for every } x .$$

A self-adjoint, nonnegative definite matrix is called a "covariance" matrix. The eigenvalues of a covariance matrix are nonnegative. Let R be a covariance matrix. Then we can define the positive square root of R by

$$\sqrt{R}\, x = \sum_1^n \sqrt{\lambda_i}\, [x, e_i] e_i , \tag{1.3}$$

where e_i are the orthonormalized eigenvectors and λ_i the corresponding eigenvalues of R. Note that

$$[Rx, x] = \|\sqrt{R}\, x\|^2$$

and in particular it follows that

$$[Rx, x] = 0$$

iff $Rx = 0$. Note also the Schwarz inequality

$$|[Rx, y]| \leq \sqrt{[Rx, x]}\, \sqrt{[Ry, y]} ,$$

which can be proved as we did for inner products by taking the minimum of the quadratic function of λ:

$$[R(x + \lambda y),\; x + \lambda y] \geq 0 .$$

Note that every covariance matrix R can be expressed as $R = AA^*$; we have only to take $A = \sqrt{R}$. Note that a "moment matrix" (of vectors) is a covariance matrix.

Functions of Matrices

Let $f(\cdot)$ be a function defined on $\mathcal{L}$ with range also in $\mathcal{L}$. It is said to be a matrix "polynomial" of degree m if

$$f(A) = \sum_0^m a_k A^k , \qquad a_m \neq 0 .$$

Let λ denote complex numbers and $f(\lambda)$ be a polynomial of degree m:

$$f(\lambda) = a_0 + a_1\lambda + \cdots + a_m\lambda^m , \qquad a_m \neq 0 .$$

Then we define

$$f(A) = a_0 I + a_1 A + \cdots + a_m A^m .$$

Thus defined, if $h(\lambda)$ is the product of the polynomials $f(\lambda)$ and $g(\lambda)$

$$h(\lambda) = f(\lambda)\, g(\lambda) ,$$

we have that

$$h(A) = f(A)g(A) = g(A)f(A) .$$

In particular, if $f(\lambda)$ is of lower degree than $p(\lambda)$, we have the "division" algorithm

$$p(\lambda) = q(\lambda)f(\lambda) + r(\lambda)$$

and correspondingly

$$p(A) = q(A)f(A) + r(A) .$$

An important result for us is the

CAYLEY-HAMILTON THEOREM. Let $\Delta(\lambda)$ denote the characteristic polynomial of A. Then

$$\Delta(A) = 0 .$$

PROOF. The proof is immediate for simple matrices. Let e_i be an eigenvector of A with eigenvalue λ_i. Then

$$\Delta(A)e_i = \Delta(\lambda_i)e_i = 0 ,$$

and since the $\{e_i\}$ is a basis for C^n it follows that the matrix $\Delta(A)$ must be zero. See one of the references listed for a proof in the more general case.

Eigenvalues of Polynomials

Let $p(\lambda)$ be defined by

$$p(\lambda) = \lambda^k + a_{k-1}\lambda^{k-1} + \cdots + a_0 .$$

Then if x is an eigenvector of A, with eigenvalue λ,

$$p(A)x = p(\lambda)x .$$

But the converse is also true. If x is an eigenvector of $p(A)$, then x is also an eigenvector of A. We can prove this readily for the case where A is simple. In that case, the eigenvectors of A form a basis and all these are eigenvectors of $p(A)$ and obviously there cannot be any more. See the references for the case where A is not simple.

Sequences

Let A_n be a sequence of matrices in $\mathcal{A}$. We say that A_n converges to a matrix A if, as $n \to \infty$,

$$\|A_n - A\| \to 0 ,$$

or

$$\|A_n - A\|_o \to 0 ,$$

or

$$\|(A_n - A)x\| \to 0$$

for every x. All these limit notions are equivalent. All of them imply (and are implied by) the statement that each entry in A_n converges to the corresponding entry in A. We note that every Cauchy sequence of matrices converges (to a matrix).

Power Series and Resolvent

Let

$$R(\lambda, A) = (\lambda I - A)^{-1} \qquad \text{for } \lambda \neq \{\lambda_i\},$$

λ_i being the eigenvalues of A. Then $R(\lambda, A)$ (called the Resolvent of A), is an analytic function of λ except for isolated poles at $\lambda = \lambda_i$. Let

$$S_n(\lambda) = \sum_0^n \frac{A^k}{\lambda^{k+1}} .$$

Then

$$\|S_{n+N}(\lambda) - S_n(\lambda)\|_o = \left\| \sum_{n+1}^{n+N} \frac{A^k}{\lambda^{k+1}} \right\|_o \leq \sum_{n+1}^{n+N} \frac{\|A^k\|_o}{|\lambda|^{k+1}} .$$

Hence $\{S_n(\lambda)\}$ is a Cauchy sequence in the operator norm if λ is such that the numerical power series

$$\sum_0^m \frac{\|A^k\|_o}{|\lambda|^k}$$

converges. Now, we know from the theory of convergence of infinite series that the numerical series converges if and only if

$$|\lambda| > r_s = \lim_{n\to\infty} \|A^n\|_o^{1/n} .$$

Let us pause briefly to show that the limit on the right exists and is finite. First of all we see that

$$\|A^n\|_o \geq r^n, \qquad \text{where } r = \max_i |\lambda_i| .$$

Hence

$$\|A^n\|_o^{1/n} \geq r .$$

Let

$$\sigma = \operatorname{Inf}_{k\geq 1} \|A^k\|_o^{1/k} .$$

Then, given $\varepsilon > 0$, we can find an integer m such that

$$\|A^m\|_o^{1/m} \leq \sigma + \varepsilon .$$

Now let us note that the function

$$f(n) = \log \|A^n\|_o$$

is subadditive

$$f(n+m) \leq f(n) + f(m) .$$

Express every integer n "modulo" m as

$$n = km + \beta, \quad 0 \leq \beta < m .$$

Then

$$\frac{f(n)}{n} = \frac{f(km+\beta)}{km+\beta} \le \frac{f(km)}{km+\beta} + \frac{f(\beta)}{km+\beta} \le \frac{f(m)}{m} + \frac{f(\beta)}{km+\beta}$$

since

$$f(km) \le kf(m) .$$

As $n \to \infty$, the second term goes to zero, so that

$$\lim_{n\to\infty} \frac{f(n)}{n} \le \sigma + \varepsilon .$$

Hence $S_n(\lambda)$ converges for $|\lambda| > \sigma$, and since

$$(\lambda I - A)S_n(\lambda) = I - \frac{A^{n+1}}{\lambda^{k+1}} ,$$

it follows that

$$\lim_{n\to\infty} S_n(\lambda) = R(\lambda, A) = \sum_0^\infty \frac{A^k}{\lambda^{n+1}} , \qquad |\lambda| > \sigma .$$

Suppose $r > \sigma$. Then for $r < |\lambda| < \sigma$ the function $R(\lambda, A)$ continues to be analytic, and hence the power series

$$\sum_0^m \frac{A^n}{\lambda^{n+1}}$$

continues to converge and hence, in particular, the nth term

$$\frac{\|A^n\|_o}{|\lambda|^{n+1}} \to 0 .$$

Hence

$$|\lambda| > \sigma ,$$

which is a contradiction. Hence

$$\log r = \lim_{n\to\infty} \frac{\log \|A^n\|_o}{n} .$$

The Exponential Function

For any real t and any A, the ('exponential' or 'power') series

$$\sum_0^\infty \frac{t^n A^n}{n!}$$

converges, since the sequence S_n of partial sums

$$S_n = \sum_0^n \frac{t^k A^k}{k!}$$

is Cauchy in the matrix norm

$$\|S_{n+p} - S\| = \left\| \sum_{n+1}^{n+p} \frac{t^k A^k}{k!} \right\| \leq \sum_{n+1}^{n+p} \frac{|t|^k \|A\|^k}{k!} ,$$

and of course

$$\sum_0^\infty \frac{|t|^k \|A\|^k}{k!} = \exp |t| \; \|A\| .$$

We define, for every t, real or complex,

$$e^{At} = \sum_0^\infty \frac{t^n A^n}{n!} . \tag{1.4}$$

Term-by-term differentiation *formally* yields

$$\frac{d}{dt} e^{At} = \sum_1^\infty \frac{t^{n-1} A^n}{n!} = A \sum_1^\infty \frac{t^{n-1} A^{n-1}}{(n-1)!} = Ae^{At} = e^{At} A .$$

To justify this rigorously, we have only to note that

$$\int_s^t \left\| Ae^{A\sigma} - \sum_0^n \frac{A^{k+1} \sigma^k}{k!} \right\| d\sigma \to 0 \qquad \text{as } n \to \infty ,$$

since the integrand goes to zero uniformly in σ. On the other hand,

$$\sum_0^n \int_s^t A^{k+1} \frac{\sigma^k}{k!} \, d\sigma = \sum_0^n A^{k+1} \frac{t^{k+1} - s^{k+1}}{(k+1)!} \to e^{At} - e^{As} .$$

Hence

$$A \int_0^t e^{As}\, ds = e^{At} - I .$$

Also we can see that

$$e^{At} e^{As} = \sum_0^\infty \sum_0^\infty A^j A^k \frac{t^j s^k}{j!k!} = \sum_0^\infty \sum_0^\infty \frac{t^j s^k}{j!k!} A^{j+k} ,$$

and since the double series converges (in norm), we can group terms, as in the "numerical" case, to obtain

$$e^{At} e^{As} = e^{A(t+s)} .$$

In particular we have that

$$(e^{At})^{-1} = e^{-At} .$$

Also

$$\|e^{At} - I\| \to 0 \qquad \text{as } t \to 0 .$$

As a consequence, we can state:

THEOREM. The differential equation

$$\frac{dx(t)}{dt} = Ax(t) , \qquad 0 < t ,$$

$$\lim_{t\to 0} x(t) = x(0) \qquad \text{given} ,$$

has the unique solution

$$x(t) = e^{At} x(0) .$$

PROOF.

$$x(t) = e^{At} x(0)$$

is clearly one solution. Let $z(t)$ be another solution. Then for every $t > 0$

$$\begin{aligned} \frac{d}{dt} e^{-At} z(t) &= -Ae^{-At} z(t) + e^{-At} Az(t) \\ &= 0 . \end{aligned}$$

Hence

$$e^{-At}z(t) = \text{constant} .$$

Taking the limit as $t \to 0+$, we see that the constant must be $x(0)$, and we must have that

$$e^{-At}z(t) = x(0) \quad \text{or} \quad z(t) = e^{At}x(0) .$$

Problem

Show that

(i) $\|e^{At}z(t) \leq e^{\|A\||t|}$; $\|e^{At} - I\| \leq (e^{\|A\|t} - 1)$, where the matrix norm may be replaced by operator norm.

(ii) Eigenvalues of $e^{At} = \{e^{\lambda_i t}\}$, where $\{\lambda_i\}$ are the eigenvalues of A.

(iii) For t real, e^{At} is a covariance if A is self-adjoint.

Calculating e^{At}

The power series form can be used for sufficiently small t when the first few terms yield an acceptable approximation. Otherwise it is necessary (in general) to know the eigenvalues λ_i. Let the characteristic polynomial $p(\lambda)$ be

$$p(\lambda) = \text{Det.}(\lambda I - A) = (\lambda - \lambda_1)^{m_1} \cdots (\lambda - \lambda_k)^{m_k} ,$$

where $\lambda_1, \ldots, \lambda_k$ are the distinct eigenvalues. Let

$$p_i(\lambda) = \frac{p(\lambda)}{(\lambda - \lambda_i)^{m_i}}$$

$$= (\lambda-\lambda_1)^{m_1} \cdots (\lambda-\lambda_{i-1})^{m_{i-1}} (\lambda-\lambda_{i+1})^{m_{i+1}} \cdots (\lambda-\lambda_k)^{m_k} ,$$

which is thus a polynomial of degree $(n - m_i)$. Expand $1/(p(\lambda))$ by partial fractions

$$\frac{1}{p(\lambda)} = \sum_{i=1}^{k} \frac{a_i(\lambda)}{(\lambda - \lambda_i)^{m_i}} ,$$

or

$$1 = \sum_{i=1}^{k} a_i(A)p_i(A) .$$

Hence, $a_i(\lambda)$, $p_i(\lambda)$ being polynomials,

$$I = \sum_{i=1}^{k} a_i(A)p_i(A) .$$

Now

$$p_i(A)(A - \lambda_i I)^{m_i} = p(A) .$$

But by the Cayley-Hamilton theorem

$$\Delta(A) = 0 .$$

Hence

$$p_i(A)(A - \lambda_i I)^{m_i} = 0 .$$

Hence, writing

$$e^{At} = e^{\lambda_i t} \sum_0^{\infty} \frac{t^n (A - \lambda_i I)^n}{n\ !}$$

we see that

$$p_i(A) e^{At} = e^{\lambda_i t} p_i(A) \sum_0^{m_i - 1} t^n \frac{(A - \lambda_i I)^n}{n!} .$$

Hence

$$\sum_{i=1}^{k} a_i(A)p_i(A) e^{At} = \sum_{i=1}^{k} a_i(A)p_i(A) e^{\lambda_i t} \sum_0^{m_i - 1} t^n \frac{(A - \lambda_i I)^n}{n!}$$

or

$$e^{At} = \sum_{i=1}^{k} a_i(A)p_i(A) e^{\lambda_i t} \sum_0^{m_i - 1} t^n \frac{(A - \lambda_i I)^n}{n!} . \tag{1.5}$$

This formula simplifies considerably if the eigenvalues $\{\lambda_i\}$ are distinct. In that case

$$e^{At} = \sum_{i=1}^{k} e^{\lambda_i t} \frac{p_i(A)}{p_i(\lambda_i)} .$$

Note that we can rewrite (1.4) as

$$e^{At} = \sum_{i=1}^{k} e^{\lambda_i t} p_i(t, A) ,$$

where $p_i(t, A)$ is a nonzero matrix polynomial function of t of degree $\leq (m_i - 1)$, for each i. Hence it follows in particular that

$$\lim_{t \to \infty} \|e^{At}\| = 0$$

if and only if $Re\ \lambda_i < 0$ for every i. We say that a matrix is "stable" if

$$\lim_{t \to \infty} \|e^{At}x\| = 0 \qquad \text{for every } x .$$

It is clear that a matrix is "stable" if and only if all its eigenvalues have negative real parts; in fact, if and only if

$$\lim_{t \to \infty} \|e^{At}\| = 0 .$$

Minimal Polynomial

Consider polynomials with leading coefficient 1, i.e., of the form

$$f(\lambda) = \lambda^k + a_{k-1}\lambda^{k-1} + \cdots + a_0 .$$

The lowest degree polynomial in this class such that

$$f(A) = 0$$

is called the minimal polynomial. By the Cayley-Hamilton theorem the lowest degree must be less than or equal to that of the characteristic polynomial. Let $\psi(\lambda)$ denote the minimal polynomial. It may well be different from the characteristic polynomial. Since for every eigenvalue λ_i we must have

$$\psi(\lambda_i) = 0 ,$$

it follows that $\psi(\lambda)$ must be a divisor of the characteristic polynomial $\Delta(\lambda)$. Let $p \leq n$ be the degree of the minimal polynomial. Then for $k \leq 0$

$$\lambda^{p+k} = q_k(\lambda)\psi(\lambda) + r_k(\lambda) ,$$

where $r_k(\lambda)$ is of degree lower than that of $\psi(\lambda)$. Hence

$$A^{p+k} = r_k(A) .$$

Hence we can write (by collecting terms appropriately)

$$e^{At} = \sum_{0}^{p-1} \phi_k(t)A^k . \tag{1.6}$$

The minimal polynomial $\psi(\lambda)$ may be calculated as follows. Let $\rho(\lambda)$ denote the characteristic polynomial. Then

$$p(\lambda) = \psi(\lambda) \times [\text{greatest common divisor of } all \text{ minors of order } (n-1) \text{ of } (\lambda I - A)] . \qquad (1.7)$$

Here is one example of a matrix whose minimal polynomial is different from the characteristic polynomial:

$$A = \begin{vmatrix} 0 & 0 & 1 & 0 \\ 0 & 0 & 0 & 1 \\ a & 0 & b & 0 \\ 0 & a & 0 & b \end{vmatrix} ,$$

$$\psi(\lambda) = (\lambda^2 - a\lambda - b) ,$$

$$\Delta(\lambda) = (\lambda^2 - a\lambda - b)^2 .$$

Similar Matrices

Two $(n \times n)$ matrices A, B are said to be similar if we can find a nonsingular matrix T such that

$$TA = BT$$

or

$$B = TAT^{-1} .$$

Note that if A is similar to B, they must have the same eigenvalues. Indeed, if

$$Ax = \lambda x ,$$

then

$$TAx = \lambda Tx ,$$

or

$$TAT^{-1}(Tx) = \lambda(Tx) ,$$

and Tx is not zero if x is not. Note also for any integer m

$$B^m = (TAT^{-1})^m = TA^m T^{-1} .$$

Hence, also

$$e^{TAT^{-1}} = Te^{At}T^{-1}$$

so that B^m is similar to A^m for every m. We have seen that every simple matrix is similar to the (any) diagonal matrix consisting of its eigenvalues.

Examples

1. Let A be a matrix such that

$$A^2 = I .$$

Then from the exponential series we can calculate that

$$\begin{aligned} e^{At} &= \sum_0^\infty \frac{A^2 t^{2k}}{(2k)!} + \sum_0^\infty \frac{A^{2k+1} t^{2k+1}}{(2k+1)!} \\ &= I\left(\sum_0^\infty \frac{t^{2k}}{(2k)!}\right) + A \sum_0^\infty \frac{t^{2k+1}}{(2k+1)!} \\ &= I \cosh t + A \sinh t . \end{aligned}$$

2.

$$A = \begin{vmatrix} -1 & -1 & 0 \\ 0 & -4 & -1 \\ 0 & 5 & 0 \end{vmatrix} ,$$

$$\begin{aligned} p(\lambda) &= (\lambda + 1)(\lambda - (-2+i))(\lambda - (-2-i)) ; \\ p_1(\lambda) &= \lambda^2 + 4\lambda + 5 ; \qquad \lambda_1 = -1 , \\ p_2(\lambda) &= (\lambda + 1)(\lambda + 2 + i) ; \qquad \lambda_2 = -2 + i , \\ p_3(\lambda) &= (\lambda + 1)(\lambda + 2 - i) ; \qquad \lambda_3 = -2 - i , \end{aligned}$$

yielding

$$a_1 = \frac{1}{2} , \qquad a_2 = \frac{i-1}{4} , \qquad a_3 = \frac{-i-1}{4} .$$

This yields

$$\begin{aligned} e^{At} = \frac{1}{2}(A_2 + 4A + 5I)e^{-t} - \frac{1}{2}(A^2 + 4A + eI)e^{-2t} \cos t \\ - \frac{1}{2}(A^2 + 2A + I)e^{-2t} \sin t . \end{aligned}$$

3. Calculate

$$e^{e^{At}}$$

for A in Example 1. We have

$$\begin{aligned} e^{e^{At}} &= I \cosh t + A \sinh t \\ &= e^{\cosh t}[I \cosh(\sinh t) + A \sinh(\sinh t)] . \end{aligned}$$

Problems

1. Calculate e^{At} for

(i)

$$A = \begin{vmatrix} -1 & -1 & 1 \\ 0 & 2 & 0 \\ -3 & -1 & 3 \end{vmatrix} .$$

(ii)

$$A = \begin{vmatrix} 0 & 1 \\ -\omega_0^2 & 0 \end{vmatrix} .$$

2. Let A, K be square matrices, and let

$$R(t) = \int_0^t e^{A\sigma} K e^{A^*\sigma} \, d\sigma .$$

Show that

$$\dot{R}(t) = AR(t) + R(t)A^* + K .$$

If K is a covariance, show that so is $R(t)$, for every t, and that $R(t)$ is monotone nondecreasing.

3. Given a polynomial

$$\lambda^n + a_{n-1}\lambda^{n-1} + \cdots + a_0$$

show that the "companion matrix"

$$\begin{vmatrix} 0 & 1 & 0 & \cdots & 0 \\ 0 & 0 & 1 & \cdots & 0 \\ \vdots & \vdots & \vdots & \ddots & \vdots \\ 0 & 0 & 0 & \cdots & 1 \\ -a_0 & -a_1 & -a_2 & \cdots & -a_{n-1} \end{vmatrix}$$

has this polynomial as its characteristic as well as minimal polynomial.

4. Find the matrix representation for the transformation of R^n into R^n given by

$$\begin{vmatrix} x_1 \\ \vdots \\ x_n \end{vmatrix} \rightarrow \begin{vmatrix} x_n \\ x_{n-1} \\ \vdots \\ x_1 \end{vmatrix} .$$

5. Find the minimal polynomials for the matrices in Problem 1.

6. Show that if A is stable:

$$\int_0^\infty e^{At}\, dt = (-A)^{-1} .$$

7. Show that the spectral radius of e^{At} is $e^{\omega_0 t}$, where

$$\omega_0 = \max_i \operatorname{Re} \lambda_i , \qquad \{\lambda_i\} = \text{eigenvalues of } A .$$

8. If A has all distinct eigenvalues $\{\lambda_i\}$, show that

$$e^{At} = \sum_0^{n-1} \phi_k(t) A^k ,$$

where

$$\begin{vmatrix} \phi_0(t) \\ \vdots \\ \phi_{n-1}(t) \end{vmatrix} = \Lambda^{-1} \begin{vmatrix} e^{\lambda_0 t} \\ \vdots \\ e^{\lambda_{n-1} t} \end{vmatrix} ,$$

$$\Lambda = \{\Lambda_{ij}\} , \qquad \Lambda_{ij} = \lambda_i^j , \qquad i, j = 0, \ldots, n-1 .$$

9. Calculate the determinant of the Vandermonde matrix:

$$\begin{vmatrix} 1 & z_1 & z_1^2 & \cdots & z_1^{p-1} \\ 1 & z_2 & z_2^2 & \cdots & z_2^{p-1} \\ \cdots & \cdots & \cdots & \cdots & \cdots \\ 1 & z_p & z_p^2 & \cdots & z_p^{p-1} \end{vmatrix}$$

10. Let $\Delta(\lambda)$ denote the characteristic polynomial of the $n \times n$ matrix A:

$$\Delta(\lambda) = \sum_0^n a_k \lambda^k .$$

Show that

$$a_0 = \text{Det. } (-A) ,$$

$$= (\text{Product of eigenvalues}) \ (-1)^n .$$

11. Show that if e_i, $i = 1,\ldots,n$ are linearly independent (column) vectors in C^n (in other words, a "basis" in C^n) the matrix

$$E = |e_1, \ldots, e_n|$$

is nonsingular.

12. Show that the moment matrix of vectors:

$$M = \{[v_i, v_j]\} , \qquad v_i \in C^n, \ i = 1,\ldots,m ,$$

is a covariance matrix.

13. Let A, B denote two simple matrices with the same characteristic polynomial. Show that they must be similar and find T such that

$$A = T^{-1} BT .$$

Show that the matrices

$$\begin{vmatrix} 0 & 1 & 0 & \cdots & 0 \\ 0 & 0 & 1 & \cdots & 0 \\ 0 & 0 & 0 & \cdots & 1 \\ \cdots & \cdots & \cdots & \cdots & \cdots \\ a_1 & a_2 & a_3 & \cdots & a_n \end{vmatrix}$$

and

$$\left|\begin{matrix} 0 & 0 & 0 & \cdots & 0 & a_1 \\ 1 & 0 & 0 & \cdots & 0 & a_2 \\ 0 & 1 & 0 & \cdots & 0 & a_3 \\ \cdots & \cdots & \cdots & \cdots & \cdots & \cdots \\ 0 & 0 & 0 & \cdots & 1 & a_n \end{matrix}\right|$$

are similar, and find the corresponding similarity transformation.

Solving Vector Equations

Let $u(t)$ be a given $(p\times 1)$ function of time, $t \geq a$, and let A be $n\times n$, and let B be $n\times p$. We want to solve the vector differential equation

$$\dot{x}(t) = Ax(t) + Bu(t)\,, \qquad t > a\,,$$

such that

$$\lim_{t\to a} x(t) = x(a) \text{ given}\,.$$

For this purpose (following the technique known for the one-dimensional version) multiply both sides by e^{-At}:

$$e^{-At}\dot{x}(t) - e^{-At}Ax(t) = e^{-At}Bu(t)\,.$$

But the left side

$$= \frac{d}{dt}(e^{-At}x(t))\,.$$

Hence

$$\frac{d}{dt}(e^{-At}x(t)) = e^{-At}Bu(t)\,.$$

Hence integrating both sides

$$\int_a^t \frac{d}{dx}(e^{-As}x(s))\,ds = \int_a^t e^{-As}Bu(s)\,ds\,.$$

Or

$$e^{-At}x(t) = e^{-Aa}x(a) + \int_a^t e^{-As}Bu(s)\,ds\,,$$

or

$$x(t) = e^{A(t-a)}x(a) + \int_a^t e^{A(t-s)}Bu(s)\,ds\,,$$

which is thus one solution. It is readily verified that

$$\lim_{t\to\infty} x(t) = x(a)\,.$$

This is only one solution. For if there were another solution, say $y(t)$, letting

$$z(t) = x(t) - y(t)\,,$$

we have that

$$\frac{d}{dt}z(t) = Az(t)\,,$$

$$\lim_{t\to\infty} z(t) = 0\,.$$

But this homogeneous equation has only the identically zero solution.

Problem

Solve (with A, B as before)

$$\dot{x}(t) = Ax(t) + Bu(t)\,, \qquad t < T\,,$$

$$\lim_{t\to T} x(t) = x(T) \text{ given.}$$

ANSWER:

$$x(t) = e^{-A(T-t)}x(T) - \int_t^T e^{A(t-s)}Bu(s)\,ds\,.$$

Rank and Range

Let A now denote an $n\times p$ rectangular matrix. We may consider it as defining a linear transformation on R^p into R^n. The rank of A is defined to be the number of linearly independent columns in the matrix A, and since the range space of A is "spanned by" (i.e., is a linear combination of) the columns of A, it follows that rank = number of linearly independent columns of A. Now for x in R^p and y in R^n we have

$$[Ax, y] = 0 \leftrightarrow [x, A^* y] = 0 .$$

Hence range of A is the orthogonal complement of the null space of A^*

$$R^n = \text{range of } A + \text{null space of } A^* . \tag{1.8}$$

Similarly:

$$R^p = \text{range of } A^* + \text{null space of } A .$$

Hence we have

$$\text{range of } A = \text{range } AA^* ,$$

$$\text{range of } A^* = \text{range } A^* A .$$

Suppose now λ_i is a nonzero eigenvalue of AA^*

$$AA^* e_i = \lambda_i e_i .$$

Then

$$A^* A(A^* e_i) = \lambda_i (A^* e_i)$$

and, since $\lambda_i \neq 0$, $A^* e_i \neq 0$, and hence $A^* e_i$ is an eigenvector of $A^* A$ with eigenvalue λ_i. Hence it follows that AA^* and $A^* A$ have the same number (including multiplicity) of nonzero eigenvalues, so that

$$\text{dim. range } AA^* = \text{dim. range } A^* A$$

and hence

$$\text{dim. range } A = \text{dim. range } A^* .$$

Hence

$$\begin{aligned} \text{rank } A &= \text{rank } A^* \\ &= \text{number of linearly independent rows (columns) of } A . \end{aligned}$$

Hence

$$\text{rank } A \leq \min (n, p) .$$

We say that A has a "full rank" if equality holds. In particular,

$$\text{rank } A = n \quad \text{iff } AA^* \text{ is nonsingular;}$$

$$\text{rank } A = p \quad \text{iff } A^* A \text{ is nonsingular.}$$

Problem

Construct a matrix A such that the range of A is *not* spanned by the eigenvectors corresponding to nonzero eigenvalues.

Hint:

$$A = \begin{vmatrix} 0 & 1 \\ -\frac{1}{4} & 1 \end{vmatrix} .$$

Representation of (Orthogonal) Projections

Let $e_1, ..., e_r$ be r linearly independent column vectors in R^n. Let P denote the (orthogonal) projection mapping R^n onto the subspace spanned by $e_1, ..., e_r$. Then by definition

$$[x - Px, e_i] = 0 , \qquad i = 1,...,r .$$

The "variational" definition is that

$$\min_{\{a_k\}} \left\| x - \sum_1^r a_k e_k \right\|^2 = \|x - Px\|^2 .$$

Let

$$Px = \sum_1^r b_i e_i .$$

Then we must have

$$[x, e_j] = \sum_1^r b_i [e_i, e_j] .$$

Let R denote the matrix with components

$$r_{ij} = [e_i, e_j] , \qquad i, j = 1,...,r ,$$

the "moment" matrix of the vectors $\{e_i\}$. Note that

$$Px = |e_1 \cdots e_r| \begin{vmatrix} b_1 \\ \vdots \\ b_r \end{vmatrix} = |e_1 \cdots e_r| \, R^{-1} \begin{vmatrix} e_1^* \\ \vdots \\ e_r^* \end{vmatrix} x \, ,$$

or the matrix representation for P is the $n \times n$ matrix

$$|e_1 \cdots e_r| \, R^{-1} \begin{vmatrix} e_1^* \\ \vdots \\ e_r^* \end{vmatrix} . \qquad (1.9)$$

The map T, mapping R^n into R^r, defined by

$$Tx = b \, , \qquad \text{where } Px = \sum_1^r b_i e_i \, ,$$

is represented by matrix $(r \times n)$

$$T = R^{-1} \begin{vmatrix} e_1^* \\ \vdots \\ e_r^* \end{vmatrix} .$$

Note that the projection has the property

$$P = P^*$$

and hence

$$[Px, \ (I - P)y] = 0$$

justifying the name "orthogonal."

Compound Matrices: Partitioning

Often we need to "partition" a large matrix into submatrices, for convenience of calculation of determinants. Conversely, we may make up a larger "compound" matrix from submatrices—"enhance" a vector by adding more components. A useful result concerning compound matrices is: the determinant of a matrix partitioned as

$$\mathcal{A} = \begin{vmatrix} 0 & B \\ C & D \end{vmatrix} ,$$

where B and C are square matrices, is given by

$$\text{Det.}\mathcal{A} = (\text{Det.}B) \cdot (\text{Det.}(-C)) . \tag{1.10}$$

If B is nonsingular, we can generalize this to

$$\text{Det.} \begin{vmatrix} A & B \\ C & D \end{vmatrix} = (\text{Det.}B) \cdot (\text{Det.}(AB^{-1}D - C)) .$$

See [1, p. 46].

Problem

Let $B_1, B_2, ..., B_m$ be $n\times p$ matrices. Let L denote the compound matrix

$$L = | B_1 \cdots B_m | .$$

Show that

$$LL^* = \sum_1^m B_i B_i^* .$$

Hence characterize the null space of L^*. What is its dimension? Specialize to $p = 1$.

Problems for Chapter 1

1. An $n\times n$ matrix $A = \{a_{ij}\}$ is said to be lower triangular if

$$a_{ij} = 0 \qquad \text{for } j > i .$$

Show that any polynomial in A is also lower triangular. Show that e^{At} is lower triangular if A is. Find the determinant of A.

2. An $n\times n$ matrix M is said to be nilpotent if

$$M^k = 0 \qquad \text{for some nonzero integer } k .$$

Show that M is nilpotent if and only if all its eigenvalues are zero. If M is nilpotent, show that $(I - M)^{-1}$ is a polynomial in M. Find e^{Mt} for M nilpotent. Show that any 2×2 nilpotent matrix must be of the form

$$\begin{vmatrix} a & b \\ c & -a \end{vmatrix},$$

where $a^2 + bc = 0$.

3. Let

$$A = \begin{vmatrix} 1 & 2 & -2 \\ -1 & 1 & 1 \\ 2 & -2 & -2 \end{vmatrix}.$$

Find e^{At}.

4. Show how to express the vector

$$\begin{vmatrix} 4 \\ 3 \\ 2 \end{vmatrix}$$

as a linear combination of the vectors

$$\begin{vmatrix} 1 \\ 0 \\ 1 \end{vmatrix}, \quad \begin{vmatrix} 0 \\ 1 \\ 1 \end{vmatrix}, \quad \begin{vmatrix} -1 \\ 1 \\ 1 \end{vmatrix}.$$

5. Show that the characteristic polynomial is also the minimal polynomial of the matrix

$$\begin{vmatrix} 0 & 0 & 0 & a_1 \\ 1 & 0 & 0 & a_2 \\ 0 & 1 & 0 & a_3 \\ 0 & 0 & 1 & a_4 \end{vmatrix}$$

and find the characteristic polynomial.

6. Calculate the minimal polynomial of the matrix

$$\begin{matrix} 0 & 0 & 0 & 1 & 0 & 0 & 0 & 0 & 0 \\ 0 & 0 & 0 & 0 & 1 & 0 & 0 & 0 & 0 \\ 0 & 0 & 0 & 0 & 0 & 1 & 0 & 0 & 0 \\ 0 & 0 & 0 & 0 & 0 & 0 & 1 & 0 & 0 \\ 0 & 0 & 0 & 0 & 0 & 0 & 0 & 1 & 0 \\ 0 & 0 & 0 & 0 & 0 & 0 & 0 & 0 & 1 \\ a_1 & 0 & 0 & a_2 & 0 & 0 & a_3 & 0 & 0 \\ 0 & a_1 & 0 & 0 & a_2 & 0 & 0 & a_3 & 0 \\ 0 & 0 & a_1 & 0 & 0 & a_2 & 0 & 0 & a_3 \end{matrix}$$

What is the characteristic polynomial?

7. Show that for any square matrix A

$$\text{Det.}(e^{At}) = e^{t(\text{Tr}\,A)} .$$

8. Generalize finding the characteristic polynomial in Problem 6 to finding the determinant of the compound matrix:

$$\begin{vmatrix} \lambda I & -I & 0 & 0 & \cdots & 0 \\ 0 & \lambda I & -I & 0 & \cdots & 0 \\ \cdots & \cdots & \cdots & \cdots & \cdots & \cdots \\ a_1 I & a_2 I & \cdots & & \cdots & a_n I \end{vmatrix} ,$$

where I is the $m \times m$ identity matrix.

Hint: Multiply the second "column" by λ, third by λ^2, ..., nth by λ^{n-1} and add to the first "column," and then use (1.10).

9. Let e_1, ..., e_n and f_1, ..., f_n be two bases (not necessarily orthonormal) in the space of $n \times 1$ column vectors C^n. Show that we can define a linear transformation by

$$Tx = \sum_1^n \alpha_i f_i ,$$

where

$$x = \sum_1^n \alpha_i e_i .$$

Show that the matrix representation for T is:

$$T = FE^{-1} ,$$

where

$$F = | f_1 \cdots f_n | ,$$

$$E = | e_1 \cdots e_n | .$$

10. Let R be a covariance matrix which has the partitioned representation

$$R = \begin{vmatrix} A_{11} & A_{12} \\ A_{21} & A_{22} \end{vmatrix} ,$$

where the submatrices are all square. Show that A_{11}, A_{22} are covariances and that

$$A_{12}^* = A_{21} .$$

11. Let A denote an $n \times n$, real-valued matrix, such that

$$A^{-1} = A^* .$$

Show that

1. Det. $A = \pm 1$;
2. A commutes with A^*;
3. the rows of A are orthogonal and orthonormal;
4. A is the identity matrix, if covariance.

Find

5. the eigenvectors and eigenvalues of A, and hence show that A is simple;
6. the trace of A;
7. the spectral radius, the matrix norm, and the operator norm;
8. the characteristic polynomial of A.

12. Show that if λ is an eigenvalue of a (square) matrix A, $\bar{\lambda}$ is an eigenvalue of A^*. Hence show that if

$$A + A^* = 0 ,$$

the eigenvalues of A are purely imaginary. What is Tr.A?

Let A denote the $(2n \times 2n)$ compound matrix

$$A = \begin{vmatrix} 0 & I \\ -D & 0 \end{vmatrix} ,$$

where I is the $n \times n$ identity and D is an $n \times n$ covariance. Show the eigenvalues are imaginary. Find det. A. Calculate e^{At} and its determinant.

13. Show that the determinant of a covariance matrix must be nonnegative.

14. Show that similar matrices must have the same characteristic polynomial.

References for Chapter 1

[1] Gantmacher, F. *Theory of Matrices*. Vol. 1. New York: Chelsea Publ. Co., 1959.

[2] Michel, A.N., and Herget, C.J. *Mathematical Foundations in Engineering and Science*. Englewood Cliffs, N.J.: Prentice-Hall Press, 1981.

[3] Padullo, L.A., and Arbib, M.A. *System Theory*. Philadelphia, PA: W.B. Saunders Co., 1974.

CHAPTER 2

SYSTEM INPUT-OUTPUT AND STATE

In this chapter we shall illustrate with examples the notions of "input-output" and "state" in describing systems.

Input-Output Relation

A system scientist deals with optimization problems in all disciplines in engineering, whether they be electrical, mechanical (including aero, hydraulic and structural), or chemical engineering. As a result, he needs to isolate those aspects which are common to all these diverse areas, even when details differ. This is particularly so today when computer simulation is an essential part of any optimization exercise. For the computer there is no difference between a truck or an airplane; the equations are similar, even if the parameters are vastly different. System theory arose in an effort to provide a unified language and approach to these problems; of optimization and simulation. Today these techniques have been extended even outside engineering; to socioeconomic problems, for example.

To an engineering student the concept of a system easiest to grasp is that of a black box with terminals hanging out; part of the terminals are "input terminals," where we may attach a voltage or current source or a torque or a force by an actuator. The other set of terminals are "output" terminals, or outputs of "sensors" that sense or measure the activity inside the black box. The idea of the black box is that one does not necessarily know what is inside; and in principle whatever is needed to optimize the performance (as expressed for example in terms of the desired output behavior) must be inferrable from the sensor data by suitable choice of "inputs." The black box (or system) is thus the totality of all the possible "input"-"output" pairs. Note that we do not necessarily assume that the "output" is determined by the "input"; indeed, the same input can give rise to several outputs. Of course, the "listing" of all possible input-output pairs can be an impossible task.

To illustrate these matters let us create a black box.

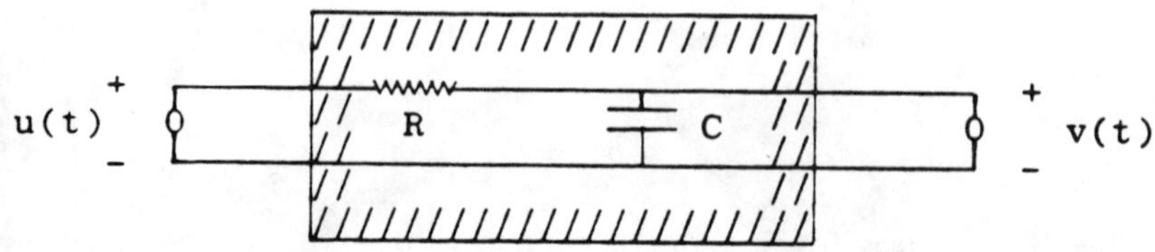

What is inside the black box is a simple R-C circuit. The input is the voltage $u(t)$, t denoting time, we can apply from a voltage source, and the output is the voltage $v(t)$ across the capacitor. Suppose we could record the voltage pairs $(u(t), v(t))$, $a < t < b$. What can we say about them in any time interval $a < t < b$? Let us take advantage of what we know is in the black box. Let $i(t)$ denote the current in the resistor. Then by Kirchhoff's Laws we know that

$$u(t) = Ri(t) + v(t) .$$

Also if $q(t)$ is the quantity of electric charge on the capacitor, we know that

$$v(t) = \frac{q(t)}{C} .$$

Here

$$\dot{q}(t) = i(t) ,$$

so that

$$i(t) = C\frac{dv}{dt} .$$

Hence any pair $(u(t), v(t))$ must satisfy

$$RC\frac{dv}{dt} + v(t) = u(t) , \quad a < t < b , \tag{2.1}$$

which we refer to as the input-output "relation." Thus the impossible task of listing all possible input-output pairs of the black box can now be replaced by saying "all possible input-output pairs satisfying the above relation"—provided of course we can show that any pair $(u(t), v(t))$ satisfying the relation is an "admissible" pair, as one we might observe at the terminals. But from the theory of differential equations we know that if

$u(\cdot)$, $v(\cdot)$ satisfy (2.1), then we can write

$$v(t) = \alpha e^{-(t-a)/RC} + \int_a^t e^{-(t-s)/RC} u(s)/RC \, ds , \qquad t \geq a , \qquad (2.2)$$

for arbitrary α. Now

$$v(a) = \alpha .$$

Hence we have only to consider the case where the charge $q(a)$ on the capacitor at time $t = a$ is given by

$$q(a) = C\alpha .$$

Then we can produce the given input-output pair $(u(t), v(t))$ at the terminals of our black box, for $a < t < b$.

Thus the black box for our purposes is "replaced" by all possible input-output pairs $(u(t), v(t))$ satisfying (2.1). Note that in this example $v(t)$ is not "determined" by $u(t)$; thus we can have the pairs

$$u(t) = 0 ; \qquad v(t) = 0 , \qquad a \leq t \leq b ,$$

$$u(t) = 0 ; \qquad v(t) = 4e^{-t/RC} , \qquad a \leq t \leq b .$$

On the other hand, the output $v(t)$ is determined if we specify the charge $q(a)$ at time $t = a$, as well as the input $u(t)$, $t \geq a$. For, we have

$$\left.\begin{aligned} q(t) &= e^{-(t-a)/RC} + \int_a^t e^{-(t-s)/RC} u(s) \, ds , \\ v(t) &= \frac{q(t)}{C} . \end{aligned}\right\} \qquad (2.2a)$$

Thus the "state" of our black box is determined by the charge of $q(\cdot)$ on the condensor, in the sense that given the state at time $t = a$ and the input for $t \geq a$, the output $v(t)$ is determined ("uniquely specified") for all $t \geq a$; and conversely, given any input-output pair $(u(t)), v(t))$, $t \geq a$, we can assign an appropriate initial charge $q(a)$ so that the pair satisfies (2.2).

State

We shall call the input-output relation such as (2.1), an "input-output description" of a system; the system being simply the totality of all input-output pairs satisfying the relation. Such a relation is "multivalued"—that is, the same input $u(\cdot)$ can give rise to many possible outputs $v(\cdot)$. The notion of "state" serves, in the first place, to remove the multivaluedness. Let us look at a slightly more complicated example than our first one to illustrate this notion more fully.

Let the system be described by the relation

$$\frac{d^2 v(t)}{dt^2} + 5\frac{dv}{dt} + 6v(t) = u(t) . \tag{2.3}$$

From the theory of differential equations we know that to specify $v(t)$ uniquely in $t \geq a$, we need to know $v(a)$ and $\dot{v}(a)$, as well as $u(t)$, $t \geq a$. Hence let $x(t)$ denote the column vector:

$$x(t) = \begin{vmatrix} v(t) \\ \dot{v}(t) \end{vmatrix} .$$

Then

$$\frac{dx(t)}{dt} = \begin{vmatrix} \dot{v}(t) \\ \ddot{v}(t) \end{vmatrix} .$$

But the right side can be expressed in terms of (as linear transformation of, and hence by a matrix multiplication of) $x(t)$

$$\begin{vmatrix} \dot{v}(t) \\ \ddot{v}(t) \end{vmatrix} = \begin{vmatrix} 0 & 1 \\ -6 & -5 \end{vmatrix} \begin{vmatrix} v(t) \\ \dot{v}(t) \end{vmatrix} + \begin{vmatrix} 0 \\ u(t) \end{vmatrix} .$$

Let

$$A = \begin{vmatrix} 0 & 1 \\ -6 & -5 \end{vmatrix} , \qquad B = \begin{vmatrix} 0 \\ 1 \end{vmatrix} .$$

Then we can express (2.3) as

$$\dot{x}(t) = Ax(t) + Bu(t) \tag{2.4}$$

and

$$v(t) = Cx(t) , \tag{2.5}$$

where C is a row vector:

$$C = | 1 \quad 0 | .$$

Thus every input-output pair satisfying (2.3) satisfies (2.4), (2.5).

Let us show now, with the help of (2.4), (2.5), that $x(t)$ can be taken as the "state" at time t. For, from Chapter 1, we know that (2.4) can be expressed as

$$x(t) = e^{A(t-a)}x(a) + \int_a^t e^{A(t-s)}Bu(s)\,ds\,, \qquad t \geq a\,, \quad (2.6)$$

and hence

$$v(t) = Ce^{A(t-a)}x(a) + \int_a^t Ce^{A(t-s)}Bu(s)\,ds\,, \quad t \geq a\,. \quad (2.7)$$

Now any $(u(\cdot), v(\cdot))$ pair satisfying (2.3), satisfies (2.4), (2.5) and hence also (2.7). Now (2.7) shows that given any $u(t)$, $t \geq a$, $v(t)$ is uniquely specified in $t \geq a$, provided we specify $x(a)$.

Let us next take any element x in R^2 and define

$$x(t) = e^{A(t-a)}x + \int_a^t e^{A(t-s)}Bu(s)\,ds\,, \qquad t \geq a\,,$$

$$v(t) = Cx(t)\,, \qquad t \geq a\,.$$

Let us show that the pair $(u(t), v(t))$ so produced satisfies relation (2.3). We can calculate

$$v(t) = Cx(t)\,,$$

$$\dot{v}(t) = C\dot{x}(t) = CAx(t) + CBu(t) = CAx(t)\,,$$

$$\ddot{v}(t) = C\ddot{x}(t) = CA\dot{x}(t) = CA^2x(t) + CABu(t)\,.$$

Hence

$$\ddot{v}(t) + 5\dot{v}t + 6v(t) = C(6I + 5A + A^2)x(t) + CABu(t)\,.$$

But

$$6I + 5A + A^2 = 0 \quad \text{(Cayley-Hamilton)}$$

and

$$CAB = 1\,,$$

or (2.3) is satisfied. Thus given any input-output pair in $t \geq a$, such a pair also satisfies (2.6), (2.7) for some $x(a)$ in R^2 and any input-output pair obtained from (2.6), (2.7) for any $x(a)$ in R^2, also satisfies (2.3). In particular, given any input $u(t)$, $t \geq a$, the corresponding output $v(t)$ is uniquely specified by specifying $x(a)$ in (2.6). Hence $x(t)$ "serves" as the state at time $t = a$. We call (2.6), (2.7) or (2.4), (2.5) a "state-space" description of the system (specified by (2.3)).

Note that in these examples the state space is a finite-dimensional linear vector space R^r and, moreover, the output is *linear* in the state for zero input and *linear* in the input for zero state. The system is thus "linear." What we have done is to deduce one state-space description—the state, at any time t, can in fact be represented differently. For example, let E be a nonsingular matrix. Let

$$y(t) = Ex(t) .$$

Then

$$\frac{d}{dt}y(t) = E\frac{dx(t)}{dt}$$

$$= EAx(t) + EBu(t) ,$$

or

$$\frac{d}{dt}y(t) = EAE^{-1}y(t) + EBu(t)$$

and

$$v(t) = CE^{-1}y(t) .$$

We have thus another state-space representation with EAE^{-1} replacing A, EB in place of B, CE^{-1} in place of C. Note that the dimension of the state space is not changed, and the new state can be obtained as a 1:1 linear transformation of the old. In particular, in a lumped-parameter linear electric circuit, the vector of initial conditions in the energy-storage elements need not be the state—any 1:1 linear transformation thereof would do as well.

Input-Output From State Description

Let us next consider the problem of deducing the input-output relation from a state-space description. Given the "state-space" description

$$\left.\begin{aligned} \dot{x}(t) &= Ax(t) + Bu(t) , \\ v(t) &= Cx(t) , \end{aligned}\right\} \qquad (2.8)$$

where A is $n \times n$, B is $n \times p$, $u(\cdot)$ is $p \times 1$, and C is $m \times n$, let us see how to deduce the input-output description of the system. Let $\psi(\cdot)$ denote the minimal polynomial of A of degree r:

$$\psi(\lambda) = \lambda^r + a_{r-1}\lambda^{r-1} + \cdots + a_0 = \sum_0^r a_k\lambda^k , \quad \text{taking } a_r = 1 .$$

We know that $r \leq n$ and that

$$\psi(A) = 0 .$$

Now

$$v(t) = Cx(t) ,$$

$$\dot{v}(t) = C\dot{x}(t) = CAx(t) + CBu(t) ,$$

$$\ddot{v}(t) = C\ddot{x}(t) = CA(Ax(t) + Bu(t)) + CB\dot{u}(t)$$

$$= CA^2x(t) + CABu(t) + CB\dot{u}(t) ,$$

and generally, using $u^k(t)$, $v^k(t)$ to denote the kth derivatives with respect to time, we have:

$$v^k(t) = CA^kx(t) + CA^{k-1}Bu(t) + \cdots + CBu^{k-1}(t) , \quad k \geq 1 ,$$

$$= CA^kx(t) + \sum_0^{k-1} CA^{k-1-j}Bu^j(t) .$$

We apply the Cayley-Hamilton theorem to obtain

$$\sum_0^r a_kCA^k = C\sum_0^r a_kA^k = C\psi(A) = 0$$

and hence, multiplying $v^k(t)$ by a_k and summing, we have:

$$\sum_0^r a_kv^k(t) = \sum_1^r a_k \sum_0^{k-1} CA^{k-1-j}Bu^j(t) ,$$

where the b_k are rectangular $(m\times p)$ matrices which can clearly be put in the form

$$\sum_0^r a_kv^k(t) = \sum_0^{r-1} b_ku^k(t) , \tag{2.9}$$

where the $\{b_k\}$ are rectangular matrices. In fact,

$$b_k = \sum_{j=k}^{r-1} a_{j+1}CA^{j-k}B .$$

Thus (2.9) is a relation that every input-output pair must satisfy—subject of course to the necessary differentiability of the input. From now on we shall assume the input has as many (continuous) derivatives as needed, and avoid what are only mathematical technicalities.

We need to show next that every input-output pair satisfying (2.9) satisfies (2.8).

EXAMPLE 2.1. Aircraft rigid-body dynamics

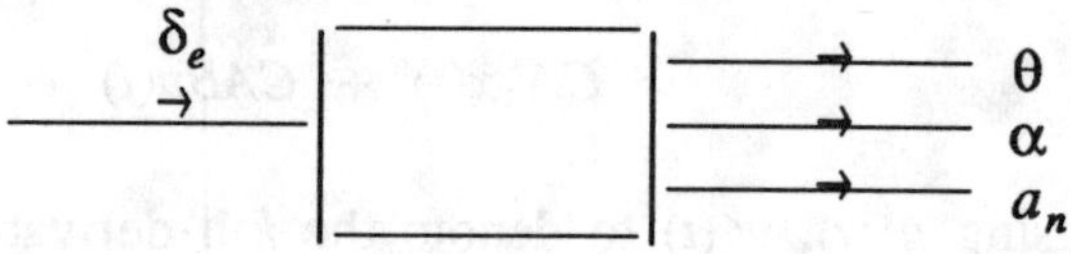

Here the input is δ_e and the outputs are θ, α, a_n with the relation

$$\begin{aligned} \ddot{\theta} &= M_\alpha \alpha + M_q \dot{\theta} + M_{\delta_e} \delta_e , \\ \dot{\alpha} &= Z_\alpha \alpha + \dot{\theta} + Z_{\delta_e} \delta_e , \\ a_n &= \frac{V}{g} (\dot{\theta} - \dot{\alpha}) . \end{aligned}$$

To obtain a state-space description define $x(t)$ as the column vector:

$$x(t) = \begin{vmatrix} \alpha(t) \\ \theta(t) \\ \dot{\theta}(t) \end{vmatrix}$$

and

$$v(t) = \begin{vmatrix} \alpha(t) \\ \theta(t) \\ a_n(t) \end{vmatrix} , \qquad u(t) = \delta_e(t) ,$$

yielding:

$$\begin{aligned} \dot{x}(t) &= Ax(t) + Bu(t) , \\ v(t) &= Cx(t) + Du(t) , \end{aligned}$$

where

$$A = \begin{vmatrix} Z_\alpha & 0 & 1 \\ 0 & 0 & 1 \\ M_\alpha & 0 & M_q \end{vmatrix},$$

$$B = \begin{vmatrix} Z_{\delta_e} \\ 0 \\ M_{\delta_e} \end{vmatrix},$$

$$C = \begin{vmatrix} 1 & 0 & 0 \\ 0 & 1 & 0 \\ -\frac{\mathrm{V}}{g} Z_\delta & 0 & 0 \end{vmatrix},$$

$$D = \begin{vmatrix} 0 \\ 0 \\ -\frac{\mathrm{V}}{g} Z_\delta \end{vmatrix}.$$

State From Input-Output Relation

Let us now consider a general input-output relation of the form (2.9) and see how to deduce an equivalent state-space description. To keep the algebra simple we shall first specialize to the single input-single output (SI-SO systems), case where both the input and output are 1×1. Thus let

$$\sum_0^r a_k \frac{d^k \mathrm{v}(t)}{dt^k} = \sum_0^r b_k \frac{d^k u(t)}{dt^k} \tag{2.10}$$

be the relation between the input $u(t)$ and the output $\mathrm{v}(t)$, both of dimension one and the $\{a_i\}$, $\{b_i\}$ are scalars. We may take $a_r = 1$. First consider the case where

$$b_k = 0, \quad k \geq 1; \qquad b_0 = 1.$$

Let

$$x(t) = \begin{vmatrix} \mathrm{v}(t) \\ \vdots \\ \mathrm{v}^{r-1}(t) \end{vmatrix} .$$

Then

$$\dot{x}(t) = \begin{vmatrix} 0 & 1 & 0 & \cdots & 0 \\ 0 & 0 & 1 & \cdots & 0 \\ \cdots & \cdots & \cdots & \cdots & \cdots \\ 0 & 0 & \cdots & 0 & 1 \\ -a_0 & -a_1 & -a_2 & \cdots & -a_{r-1} \end{vmatrix} \begin{vmatrix} \mathrm{v}(t) \\ \vdots \\ \mathrm{v}^{r-1}(t) \end{vmatrix} + \begin{vmatrix} 0 \\ 0 \\ \vdots \\ 1 \end{vmatrix} u(t) .$$

Let A denote the $r \times r$ matrix

$$A = \begin{vmatrix} 0 & 1 & 0 & \cdots & 0 \\ 0 & 0 & 1 & \cdots & 0 \\ \cdots & \cdots & \cdots & \cdots & \cdots \\ 0 & 0 & 0 & \cdots & 1 \\ -a_0 & -a_1 & -a_2 & \cdots & -a_{r-1} \end{vmatrix} \tag{2.11}$$

The minimal polynomial of this polynomial is the same as the characteristic polynomial. Let

$$B = \begin{vmatrix} 0 \\ 0 \\ 0 \\ \vdots \\ 0 \\ 1 \end{vmatrix} , \qquad C = | 1 \; 0 \; 0 \; 0 \; \cdots \; 0 | , \tag{2.12}$$

B is $r \times 1$; C is $1 \times r$. Then we have the state-space representation

$$\left. \begin{aligned} \mathrm{v}(t) &= Cx(t) , \\ \dot{x}(t) &= Ax(t) + Bu(t) . \end{aligned} \right\} \tag{2.13}$$

Next let us allow $\{b_k\}$ to be nonzero. Then replacing $u(t)$ by the right side of (2.10), we have:

$$\mathrm{v}(t) = Cx(t) ,$$

$$\dot{x}(t) = Ax(t) + B \sum_0^r b_k \frac{d^k u(t)}{dt^k} ,$$

or

$$x(t) = e^{At}x(0) + \int_0^t e^{A(t-s)} B\left[\sum_0^r b_k \frac{d^k u(s)}{ds^k}\right] ds\,, \qquad t \geq 0\,.$$

Let $Bb_k = B_k$. Integrating by parts:

$$\int_0^t e^{A(t-s)} B_k \frac{d^k u(s)}{ds^k} = {}_0^t[e^{A(t-s)}B_k u^{k-1}(s)] + \int_0^t e^{A(t-s)} A B_k u^{k-1}(s)\, ds$$

$$= B_k u^{k-1}(t) - e^{At}B_k u^{k-1}(0) + \int_0^t e^{A(t-s)} A B_k u^{k-1}(s)\, ds$$

$$= (B_k u^{k-1}(t) + AB_k u^{k-2}(t) + \cdots + A^{k-1} B_k u(t))$$

$$- e^{At}(B_k u^{k-1}(0) + AB_k u^{k-1}(0) + \cdots + A^{k-1} B_k u(0))$$

$$+ \int_0^t e^{A(t-s)} A^k B_k u(s)\, ds$$

$$= \sum_0^{k-1} A^j B_k u^{k-1-j}(t) - e^{At} \sum_0^{k-1} A^j B_k u^{k-1-j}(0)$$

$$+ \int_0^t e^{A(t-s)} A^k B_k u(s)\, ds\,.$$

Hence

$$x(t) = e^{At}\left(x(0) - \sum_{k=0}^{r} \sum_{j=0}^{k-1} A^j Bb_k u^{k-1-j}(0)\right) + \sum_{k=1}^{r} \sum_{j=0}^{k-1} A^j Bb_k u^{k-1-j}(t)$$

$$+ \int_0^t e^{A(t-s)} \left[\sum_0^r A^k Bb_k\right] u(s)\, ds\,.$$

Hence let

$$y(t) = x(t) - \sum_{k=0}^{r} \sum_{j=0}^{k-1} A^j Bb_k u^{k-1-j}(t)\,.$$

Then we have

$$y(t) = e^{At}y(0) + \int_0^t e^{A(t-s)} \tilde{B}u(s)\, ds ,$$

where

$$\tilde{B} = \sum_0^r A^k B b_k .$$

Moreover,

$$v(t) = Cx(t) = Cy(t) + C \sum_0^r \sum_0^{k-1} A_j B b_k u^{k-1-j}(t) .$$

Now for our C, A and B, we note that

$$CA^jB = 0 , \qquad j = 0, 1, ..., r-2 ,$$

$$CA^{r-1}B = 1 .$$

Hence

$$C \sum_{k=0}^{r} \sum_{j=0}^{k-1} A^j B_k u^{k-1-j}(t) = CA^{r-1}B_r u(t) .$$

Hence we have state-space representation:

$$\left. \begin{aligned} v(t) &= Cy(t) + Du(t) , \\ \dot{y}(t) &= Ay(t) + \tilde{B}u(t) , \end{aligned} \right\} \tag{2.14}$$

where

$$D = CA^{r-1}B_r = b_r .$$

Let us next go on to consider the general case (2.9) where now the output $v(\cdot)$ is $m\times 1$ and the input $u(\cdot)$ is $p\times 1$ and the $\{a_k\}$ are scalars and the $\{b_k\}$ are $m\times p$ matrices. The only changes to be made in the arguments are in the notation, to allow for the increase in dimension. The new matrices will all be "compound" matrices with (sub) matrices as components. Thus A is the compound matrix

$$A = \begin{vmatrix} 0 & I & 0 & \cdots & 0 \\ 0 & 0 & I & \cdots & 0 \\ \cdots & \cdots & \cdots & \cdots & \cdots \\ 0 & 0 & 0 & \cdots & I \\ -a_0 I & -a_1 I & -a_2 I & \cdots & -a_{r-1} I \end{vmatrix} ,$$

where I is the $m \times m$ identity matrix, so that A itself is $(mr \times mr)$. We again consider first the special case

$$\sum_0^r a_k \frac{d^k \mathrm{v}(t)}{dt^k} = u(t) , \tag{2.15}$$

where the input $u(\cdot)$ is $m \times 1$. We define B as the compound matrix:

$$B = \begin{vmatrix} 0 \\ 0 \\ \vdots \\ I \end{vmatrix} ,$$

with each component matrix being $m \times m$ and there are r of them, so that B is $mr \times m$. In a similar way, we define C as the compound matrix:

$$C = | I \ 0 \ 0 \ \cdots \ 0 | ,$$

where I is $m \times m$, and there are r of them so that C is $m \times mr$. Then we have

$$\dot{x}(t) = Ax(t) + Bu(t) \tag{2.13a}$$

and

$$\mathrm{v}(t) = Cx(t)$$

analogous to (2.13). We next allow the $\{b_k\}$ to be nonzero. Then putting

$$\sum_0^r b_k u^k(t)$$

in place of $u(t)$, we have

$$\dot{x}(t) = Ax(t) + B\left[\sum_0^r b_k u^k(t)\right] ,$$

"solving" which we have

$$x(t) = e^{At}x(0) + \int_0^t e^{A(t-s)} B\left[\sum_0^r b_k u^k(s)\right] ds .$$

We can proceed exactly as before, and again defining

$$\tilde{B} = \sum_0^r A^k B b_k$$

and defining $y(t)$ as before, and noting that with the new definitions we still have:

$$CA^j B = 0 , \qquad j = 0, \ldots, r-2 ,$$

$$CA^{r-1} B = I ,$$

where I is the $m \times m$ Identity matrix, we obtain exactly (2.14) again, as before.

The state-space representation (2.14) is said to be "proper" if $D = 0$. Note that $D = 0$ if $b_r = 0$ in the input-output relation. If we associate the polynomial

$$D(\lambda) = \sum_0^r a_k \lambda^k$$

with the left side of (2.9) (called the denominator polynomial), and the associated matrix polynomial

$$N(\lambda) = \sum_0^r b_k \lambda^k$$

with the right side (called the numerator polynomial), the state space is proper if the degree of the numerator polynomial is less than that of the denominator polynomial.

Let us show that the two descriptions (2.9) and (2.14) are equivalent. From the way in which we constructed (2.14) we see that any $(u(\cdot), v(\cdot))$ pair that satisfies (2.10) satisfies (2.14). Let us go the other way. Suppose that we consider a $(u(\cdot), v(\cdot))$ pair that satisfies (2.14). First of all we note the minimal polynomial of A is

$$\psi(\lambda) = \sum_0^r a_k \lambda^k , \qquad a_r = 1 \tag{2.15}$$

(the characteristic polynomial is $(\psi(\lambda))^m$; see Chapter 1). Asssuming the necessary differentiability for $u(t)$ and differentiating the output-state equation in (2.14) we see that

$$\sum_0^r a_k \frac{dv^k(t)}{dt^k} = \left[\sum_0^r a_k C A^k\right] y(t) + \sum_1^r a_k C \sum_{j=0}^{k-1} A^{k-1-j} \tilde{B} \frac{du^j(t)}{dt^j} . \tag{2.16}$$

The first term on the right in (2.16) is zero because of (2.15). The second term on the right in (2.16), upon changing the order of summation, yields:

$$C \sum_{j=0}^{r-1} \left[\sum_{m=j}^{r-1} a_{m+1} A^{m-j} \tilde{B}\right] u^j(t) .$$

Hence we obtain

$$\sum_0^r a_k v^k(t) = \sum_0^{r-1} \left[C \sum_{m=j}^{r-1} a_{m+1} A^{m-j} \tilde{B} + a_j D \right] u^j(t) + D u^r(t) , \quad (2.17)$$

and

$$D = b_r .$$

Note that (2.17) is already in the general form as (2.10), except for identifying the coefficients of $u^j(t)$, for $j \le r-1$. Thus we need to show that

$$C \sum_{m=j}^{r-1} a_{m+1} A^{m-j} \tilde{B} + a_j D_r = b_j , \quad j \le r-1 . \quad (2.18)$$

We can express the left side as

$$\sum_{m=j}^{r-1} a_{m+1} C A^{m-j} \sum_0^r A^k B b_k + a_j b_r$$

$$= \sum_{k=0}^{n} \left[\sum_{m=j}^{r-1} a_{m+1} (C A^{m-j+k} B) \right] b_k + a_j b_r .$$

Now the tools available are:

$$(CA^p B) b_k = 0 \qquad \text{for } p \le r-2 , \quad (2.19)$$

$$= b_k \qquad \text{for } p = r-1 , \quad (2.20)$$

and the Cayley-Hamilton theorem. Using (2.19) it follows that

$$\sum_{m=j}^{r-1} a_{m+1} (C A^{m-j+k} B) b_k = 0 \qquad \text{for } k \le j-1 ,$$

and for $k = j$, it is

$$= \sum_{m=j}^{r-1} a_{m+1} (C A^m B) b_j$$

$$= \sum_{m=j+1}^{r-1} a_m (C A^{m-1} B) b_j$$

$$= a_r (C A^{k-1} B) b_j$$

$$= a_r b_j \qquad \text{(using (2.20))}$$

$$= b_j \qquad \text{(since } a_r = 1\text{)} .$$

For $k = j + p, \ \ 1 \le p \le k - j$, it is

$$= \sum_{m=j}^{r-1} a_{m+1}(CA^{m+p}B)b_{j+p}$$

$$= \sum_{m=j}^{r} a_m(CA^{m+p-1}B)b_{j+p} ,$$

which, using the Cayley-Hamilton theorem,

$$\sum_{j+1}^{r} a_m A^m = -\sum_{0}^{j} a_m A^m ,$$

becomes:

$$-\sum_{m=0}^{j} a_m (CA^{m+p-1}B)b_{j+p} .$$

Now in this sum,

$$m + p - 1 \le j + p - 1 \le r - 1$$

and hence

$$(CA^{m+p-1}B) = 0$$

unless

$$m + p - 1 = k - 1$$

or

$$p = r - j ; \qquad m = j ,$$

and in that case, it is

$$= -a_j b_r .$$

Hence

$$\sum_{k=0}^{r} \left[\sum_{m=j}^{r-1} a_{m+1}(CA^{m-j+k}B) \right] b_k = b_j - a_j b_r$$

and hence (2.18) follows. Hence (2.16) is the same as (2.9).

Starting with the input-output relation (2.9) describing the system we have thus arrived at an equivalent state-space description of the system (2.14). The state-space description has the "generic" form:

$$\left.\begin{aligned} \dot{x}(t) &= Ax(t) + Bu(t)\,, \\ v(t) &= Cx(t) + Du(t)\,. \end{aligned}\right\}. \tag{2.21}$$

However, starting from (2.9), we have *not* shown that we can get back to (2.8). We have two state-space representations with the same input-output relation. We defer the resolution of this question to Chapter 4. Note in this connection that given a state-space representation such as (2.21), the matrices A, B, C are *not* unique. We can get other equivalent state-space representations with different A, B, C. Thus let T be any nonsingular $n \times n$ matrix. Define

$$z(t) = Tx(t)\,, \tag{2.22}$$

where $x(\cdot)$ is the solution to the state equation in (2.22). Then

$$\dot{z}(t) = T\dot{x}(t) = TA(t) + TBu(t) = TAT^{-1}z(t) + TBu(t)$$

and

$$v(t) = CT^{-1}(t) + Du(t)\,.$$

The system

$$\begin{aligned} \dot{z}(t) &= Fz(t) + Gu(t)\,, \\ v(t) &= Hz(t) + Du(t)\,, \end{aligned} \tag{2.23}$$

where

$$F = TAT^{-1}\,; \qquad G = TB\,; \qquad H = CT^{-1}\,,$$

is clearly equivalent to (2.21) in the sense that the set of input-output pairs is the same. Note that F is similar to A.

In obtaining an input-output relation (2.9) from the state-space description (2.8), we proceeded by taking the minimal polynomial $\psi(\lambda)$ of A, but the only property we really used was that

$$\psi(A) = 0\,.$$

Let us therefore take any polynomial $p(\lambda)$:

$$p(\lambda) = \sum_0^n p_k \lambda^k \tag{2.25}$$

such that

$$p(A) = 0 \, .$$

Then all our arguments would go through and we would obtain an input-output relation of the form

$$\sum_0^n p_k \mathrm{v}^k(t) = \sum_0^{n-1} d_k u^k(t) \, , \tag{2.26}$$

which is not equivalent to (2.9). Before we examine the general case let us look at a simple example. Let

$$\dot{x} = Ax + Bu \, ,$$

$$\mathrm{v} = Cx \, ,$$

where

$$A = \begin{vmatrix} -1 & 0 \\ 0 & -1 \end{vmatrix} , \qquad B = \begin{vmatrix} 1 \\ 1 \end{vmatrix} , \qquad C = | 1 \;\; 1 | \, ,$$

so that we have a SI-SO system. Let us take the characteristic polynomial:

$$p(\lambda) = (\lambda + 1)^2 \, .$$

Then we would get the input-output relation:

$$\frac{d^2\mathrm{v}}{dt^2} + 2\frac{d\mathrm{v}}{dt} + \mathrm{v}(t) = 2u(t) + 2u'(t) \, . \tag{2.27}$$

On the other hand, if we took the minimal polynomial of A:

$$\psi(\lambda) = \lambda + 1 \, ,$$

we would have

$$\frac{d\mathrm{v}}{dt} + \mathrm{v}(t) = 2u(t) \, , \tag{2.28}$$

which is at once a state-space (taking $\mathrm{v}(t)$ to be the state) representation. The crucial point involved here is that any $(u(\cdot), \mathrm{v}(\cdot))$ pair that satisfies (2.28) does satisfy (2.27), but *not* conversely. To see this let $(u(\cdot), \mathrm{v}(\cdot))$ satisfy (2.27). Define

$$w(t) = \frac{d\mathrm{v}}{dt} + \mathrm{v}(t) - 2u(t) \, ,$$

which is the difference between the two sides of (2.28). Then (2.27) becomes

$$\frac{dw(t)}{dt} + w(t) = 0 .$$

But this equation has, for example, the solution:

$$w(t) = e^{-t} w(0) , \qquad t \geq 0 .$$

In particular $w(t)$ need not be zero, and hence (2.28) need not be satisfied.

Let us now get back to the general case of (2.26). The main point is that since

$$p(A) = 0 ,$$

the minimal polynomial $\psi(\lambda)$ must divide $p(\lambda)$. Hence let

$$q(\lambda) = \frac{p(\lambda)}{\psi(\lambda)} = \sum_0^m q_k \lambda^k$$

for appropriate m. Consider the input

$$u(t) = u e^{\lambda t} .$$

Then the pair

$$u(t) = u e^{\lambda t} ,$$

$$\mathrm{v}(t) = \frac{\left(\sum_0^{r-1} b_k \lambda^k \right) u e^{\lambda t}}{\psi(\lambda)}$$

for every λ excepting the finite number of roots of $\psi(\lambda)$. This satisfies (2.8) and relation (2.14) and hence also (2.26). Hence we must have

$$\left(\sum_0^n p_k \lambda^k \right) \frac{\left(\sum_0^{r-1} b_k \lambda^k \right)}{\psi(\lambda)} u = \left(\sum_0^{n-1} d_k \lambda^k \right) u$$

for all u and hence

$$q(\lambda) = \sum_0^{r-1} b_k \lambda^k \bullet \sum_0^{n-1} d_k \lambda^k . \tag{2.29}$$

Let D denote the differential operator d/dt. Then

$$q(D)u = \sum_0^{r-1} b_k D^k \bullet \sum_0^{n-1} d_k D^k u$$

for any $u(\cdot)$. Let $(u(\cdot), \mathrm{v}(\cdot))$ be an input-output pair satisfying (2.26). Then

$$p(D)\mathrm{v} = q(D)[\psi(D)\mathrm{v}] = \sum_0^{n-1} d_k D^k u = \sum q(D)\left[\sum_0^{r-1} b_k D^k u\right].$$

Hence

$$q(D)w = \sum_0^m q_k \frac{d^k w(t)}{dt^k} = 0\,, \tag{2.30}$$

where

$$w(t) = \psi(D)\mathrm{v}(t) - \sum_0^{r-1} b_k D^k u(t)\,,$$

which is the difference between the left and right sides of (2.9). For $(u(\cdot), \mathrm{v}(\cdot))$ to satisfy (2.9) we must have $w(t)$ identically zero. But we can clearly find nonzero solutions of (2.30). Hence there are input-output pairs which satisfy (2.26) but not (2.9), as we sought to prove.

We may note that if we follow our procedure for deducing a state-space description from (2.26), it would of course be different from that from (2.9), and in particular the state-space dimension would be larger. What distinguishes these state spaces involves the notion of controllability, which is the subject of our next chapter.

Problems for Chapter 2

1. Given the SI-SO input-output relation (2.3) in the text:

$$\frac{d^2\mathrm{v}}{dt^2} + 6\frac{d\mathrm{v}}{dt} + 8\mathrm{v}(t) = u(t)\,,$$

find a state-space description, now taking

$$x(t) = \begin{vmatrix} \mathrm{v}(t) + \dot{\mathrm{v}}(t) \\ \mathrm{v}(t) - \dot{\mathrm{v}}(t) \end{vmatrix}\,,$$

either directly or by an appropriate 1:1 transformation on the previous state definition. Let

$$u(t) = e^{6t}\,, \qquad t \geq 0\,,$$

$$\mathrm{v}(t) = ae^{6t}\,, \qquad t \geq 0\,.$$

For what value of a will this be an admissible input-output pair for this system, and what is the corresponding state at $t = 0$? For any $t > 0$?

2. Given the SI-SO input-output relation:

$$\frac{d^2 v}{dt^2} + 6\frac{dv}{dt} + 8v(t) = \frac{du}{dt} + 3u(t) ,$$

find an equivalent state-space description.

3. Deduce an equivalent input-output relation from the given state-space description in Example 2.1. What is the relationship of the input-output relation you derive to the original?

4. Given the state-space representation

$$\dot{x}(t) = Ax(t) + Bu(t) ,$$

$$v(t) = Cx(t) ,$$

where A is the $r \times r$ matrix

$$A = \begin{vmatrix} 0 & 0 & 0 & \cdots & 0 & -a_0 \\ 1 & 0 & 0 & \cdots & 0 & -a_1 \\ 0 & 1 & 0 & \cdots & 0 & -a_2 \\ \cdots & \cdots & \cdots & \cdots & \cdots & \cdots \\ 0 & 0 & 0 & \cdots & 1 & -a_{r-1} \end{vmatrix} ,$$

B is $r \times 1$:

$$B = \begin{vmatrix} 1 \\ 0 \\ \vdots \\ 0 \end{vmatrix} ,$$

and C is $1 \times r$:

$$C = | 0 \; 0 \; \cdots \; 1 | ,$$

find the equivalent input-output relation.

5. Given an input-output relation (2.9) we may differentiate both sides with respect to t and obtain new input-output relations. For a specific example take the SI-SO case:

$$\frac{d^2 v}{dt^2} + 6\frac{dv}{dt} + 8v(t) = u(t) .$$

Differentiating both sides once we obtain

$$\frac{d^3\text{v}}{dt^3} + 6\,\frac{d^2\text{v}}{dt^2}\;\;8\,\frac{d\text{v}}{dt} = \frac{du(t)}{dt}\,.$$

Show that in this case we have created new input-output pairs which do not satisfy the older input-output relation. Derive an equivalent state space for each and discuss the difference.

6. In Problem 4 define a new state by the 1:1 transformation

$$y = Tx\,,$$

where T is $r \times r$:

$$T = \left|\begin{matrix} 0 & 0 & 0 & \cdots & 0 & 1 \\ 0 & 0 & 0 & \cdots & 1 & 0 \\ \cdots & \cdots & \cdots & \cdots & \cdots & \cdots \\ 0 & 1 & 0 & \cdots & 0 & 0 \\ 1 & 0 & 0 & \cdots & 0 & 0 \end{matrix}\right|\,.$$

Find the new state-space representation.

7. In Problem 4, assume that the characteristic polynomial

$$\sum_0^r a_k \lambda^k$$

has distinct roots. Find a nonsingular matrix E which diagonalizes A, such that EAE^{-1} is diagonal. Define

$$y = Ex$$

and find the new state-space representation. As a concrete example, take

$$A = \left|\begin{matrix} 0 & +5 \\ 1 & -6 \end{matrix}\right|$$

and determine E explicitly.

8. Given the state-space representation in Problem 4, define the 1:1 transformation

$$y = Ex$$

by

$$y = \sum_{0}^{r-1} x_k A^k B \,, \qquad x = \begin{vmatrix} x_0 \\ \vdots \\ x_{r-1} \end{vmatrix} .$$

Find

$$EAE^{-1} \quad \text{and} \quad EB$$

and the state-space representation corresponding to the transformation.

CHAPTER 3

LINEAR DYNAMIC SYSTEMS: CONTROLLABILITY

We shall call a system with a given state-space description

$$\begin{aligned} \dot{x}(t) &= Ax(t) + Bu(t) , \\ v(t) &= Cx(t) + Du(t) , \end{aligned} \tag{3.0}$$

where A, B, C, D are constant matrices of arbitrary (but finite) dimensions, a "time-invariant linear dynamic system." We shall now consider three important structural features of such systems that play a crucial role in problems of filtering and control: controllability, observability and stability. The first two serve, in particular, to sort out the seeming myriad of state-space and input-output descriptions. This chapter treats Controllability.

Controllability

The first notion is that of "Controllability" (actually the term "reachability" which is also sometimes used, is more appropriate since controllability need not have anything to do with "controlling" the system).

DEFINITION 3.1. We shall say that a (given) state space is controllable if we can reach any state in some time, starting from the zero state, by applying a suitable input $u(\cdot)$.

By definition, if the state space is controllable, given any state x, we can express it as

$$x = \int_0^t e^{A(t-\sigma)} Bu(\sigma)\, d\sigma$$

for some $u(\cdot)$ and some t. Let us now deduce some tests for determining controllability. First of all, for each t, let $\Omega(t)$ denote the class of states reachable in time interval t, starting from the zero state. Then, if n is the dimension of the state space (A is $n \times n$), controllability implies

$$R^n = \bigcup_{t>0} \Omega(t) . \tag{3.1}$$

Let E be any set in R^n. By

$$[x, E] = 0 \qquad [\text{alternately } x \perp E]$$

we mean that x is orthogonal to every element of E:

$$[x, z] = 0 \quad \text{for every } z \text{ in } E .$$

Note now the obvious fact that

$$[x, R^n] = 0$$

implies that x must be zero. Hence (3.1) is equivalent to saying that: if for some x in R^n,

$$[x, \Omega(t)] = 0 \quad \text{for every } t > 0 ,$$

then x must be zero.

Let us carry this a step further. Let $x \in R^n$. Let z be an element in $\Omega(t)$. Then

$$[x, z] = \left[x, \int_0^t e^{A(t-\sigma)} Bu(\sigma)\, d\sigma \right]$$

for some input $u(\cdot)$. Make a change of variable in the integrand and write

$$\int_0^t e^{A(t-\sigma)} Bu(\sigma)\, d\sigma = \int_0^t e^{As} B\tilde{u}(s)\, ds ,$$

where

$$\tilde{u}(s) = u(t-s) .$$

Hence we can write:

$$[x, z] = \int_0^t [B^* e^{A^* s} x, \ \tilde{u}(s)]\, ds . \tag{3.2}$$

Let us now choose

$$\tilde{u}(s) = B^* e^{A^* s} x ,$$

or, in other words, choose the input $u(\cdot)$ defined by

$$u(\sigma) = B^* e^{A^*(t-\sigma)} x , \qquad 0 \le \sigma \le t .$$

Then

$$[x, \Omega(t)] = 0 \tag{3.3}$$

must imply that

$$\int_0^t [B^* e^{A^* s} x,\ B^* e^{A^* s} x]\, ds = 0$$

or

$$B^* e^{A^* s} x = 0 , \qquad 0 \le s \le t . \tag{3.4}$$

Conversely, if (3.4) holds, then (3.2) is zero for every z in $\Omega(t)$, or (3.3) follows. Multiplying on the left by $e^{As}B$ in (3.4), we have

$$e^{As} B B^* e^{A^* s} x = 0 ,$$

so that

$$\left(\int_0^t e^{As} B B^* e^{A^* s}\, ds \right) x = 0 .$$

Let

$$R(t) = \int_0^t e^{As} B B^* e^{A^* s}\, ds , \qquad 0 < t . \tag{3.5}$$

Then $R(t)$ is a covariance for each t, and

$$[x, \Omega(t)] = 0 \iff R(t)x = 0 , \tag{3.6}$$

since

$$R(t)x = 0 \iff [R(t)x, x] = 0 ,$$
$$\iff B^* e^{A^* s} x = 0 , \qquad 0 \le s \le t .$$

Hence we can state:

THEOREM 3.1. The state space is controllable if and only if $R(t)x = 0$ for every $t > 0$ implies that $x = 0$.

PROOF. Suppose that $R(t)x = 0$ for *every* $t > 0$ implies that $x = 0$. Suppose that state space is not controllable. Then we can find *nonzero* x such that

$$\left[x, \bigcup_{t>0} \Omega(t)\right] = 0 .$$

In particular

$$[x, \Omega(t)] = 0 \qquad \text{for every } t > 0 .$$

Then, as we have seen,

$$R(t)x = 0 \qquad \text{for every } t > 0 ,$$

which contradicts the hypothesis.

Suppose conversely the state space is controllable. Suppose contrariwise for some nonzero x,

$$R(t)x = 0 \qquad \text{for every } t > 0 .$$

Then, as we have seen,

$$[x, \Omega(t)] = 0 \qquad \text{for every } t > 0 .$$

Hence the state space is not controllable, contradicting the hypothesis.

THEOREM 3.2. The state space is controllable if and only if

$$B^* e^{A^* s} x = 0 , \qquad 0 \leq s < \infty ,$$

implies that $x = 0$.

PROOF. The proof is immediate since

$$R(t)x = 0$$

for every t if and only if

$$B^* e^{A^* s} x = 0 , \quad 0 \leq s < \infty .$$

These results can be strengthened considerably. For this purpose let us explore further the properties of $R(t)$ and $\Omega(t)$. For any x in R^n, define

$$u(\sigma) = B^* e^{A^*(t-\sigma)} x , \qquad 0 \leq \sigma \leq t .$$

Then

$$R(t)x = \int_0^t e^{A(t-\sigma)} B B^* e^{A^*(t-\sigma)} x \, d\sigma$$

$$= \int_0^t e^{A(t-\sigma)} Bu(\sigma) \, d\sigma , \qquad \text{for each } t > 0 ,$$

from which it follows that the range of $R(t)$ is contained in $\Omega(t)$ for each $t > 0$.

Let us next exploit (for the first time) that the state space is finite-dimensional.

THEOREM 3.3.

$$\Omega(t) = \text{range of } R(t) .$$

PROOF. It only remains to prove that $\Omega(t)$ is contained in the range of $R(t)$.

Let us prove this by an elementary argument, which will also lead us to another and more constructive test for controllability. For this purpose we invoke the Cayley-Hamilton theorem and write

$$e^{At} = \sum_0^{n-1} \phi_k(t) A^k .$$

Hence

$$\int_0^t e^{As} Bu(s) \, ds = \sum_0^{n-1} A^k B \left[\int_0^t \phi_k(s) u(s) \, ds \right] = \sum_0^{n} A^k B u_k(t) ,$$

where

$$u_k(t) = \int_0^t \phi_k(s) u(s) \, ds .$$

Consider now the compound matrix ($n \times np$, since B is $n \times p$)

$$L_c = | B \; AB \; A^2B \cdots A^{n-1}B | . \tag{3.7}$$

We can consider this as defining a linear transformation on $R^p \times R^p \times \cdots \times R^p$, n-times (or R^{np}) into R^n. Let us call L_c the "controllability" matrix. It is clear that $\Omega(t)$ for every t is actually contained in the range of L_c. Now L_c^* denoting the adjoint of L_c, we have

$$L_c L_c^* = \sum_0^{n-1} A^k B B^* A^{*k} .$$

On the other hand,

$$R(t) = \sum_0^{n-1} \sum_0^{n-1} r_{jk}(t) A^j B\, B^* A^{*k} , \tag{3.8}$$

where

$$r_{jk}(t) = \int_0^t \phi_j(s) \phi_k(s)\, ds .$$

We shall now prove that the null spaces of $L_c L_c^*$ and $R(t)$ are the same. Suppose that

$$R(t)x = 0 .$$

Then we know that

$$B^* e^{A^* s} x = 0 , \qquad 0 \leq s \leq t .$$

Hence

$$\frac{d^k}{ds^k} (B^* e^{A^* s} x) \Big|_{s=0} = 0 \qquad \text{for every } k ,$$

or

$$B^* A^{*k} x = 0 \qquad \text{for every } k .$$

Hence

$$L_c L_c^* x = 0 .$$

Suppose

$$L_c L_c^* x = 0 .$$

Then

$$[L_c L_c^* x, \ x] = \sum_0^{n-1} [A^k B^* A^{*k} x, \ x] = \sum_0^{n-1} \| B^* A_k^* x \|^2 = 0 ,$$

or

$$B^* A^{*k} x = 0 , \qquad k = 0, 1, ..., n-1 .$$

Hence

$$\sum_0^{n-1} \phi_k(s) B^* A^{*k} x = B^* e^{A^* s} x = 0 , \qquad 0 \leq s ,$$

and consequently

$$R(t)x = 0 \qquad \text{for every } t .$$

But the range of a self-adjoint matrix is the orthogonal complement of the null space. Hence both $L_c L_c^*$ and $R(t)$ have the same range space. But $\Omega(t)$ is contained in the range of L_c and $L_c L_c^*$ and is therefore also of $R(t)$. Hence we have

$$\Omega(t) = \text{range of } R(t) = \text{range of } L_c L_c^* = \bigcup_{t>0} \Omega(t) . \qquad (3.9)$$

Immediate from this is a constructive test for controllability.

THEOREM 3.4. The state space is controllable if and only if the rank of the controllability matrix L_c is equal to n (or, L_c has "full" rank).

PROOF. Since

$$\bigcup_{t>0} \Omega(t) = \text{range } L_c L_c^* ,$$

it follows that the state space is controllable *iff* $L_c L_c^*$ is nonsingular, or L_c has full rank n. Q.E.D.

Note that the subspaces $\Omega(t)$ all have the same dimension for every $t > 0$, being equal to the rank of L_c. Also if $R(t)$ is nonsingular for some $t > 0$, then it is nonsingular for every $t > 0$. The state space is controllable iff $R(t)$ is nonsingular for some $t > 0$. Also

$$R^* e^{A^* s} x = 0 , \qquad 0 \leq s \leq t ,$$

for *some* $t > 0$, implies the same for every $t > 0$.

Suppose the state space is controllable. Then it is possible, by definition, to go from the zero state to any state in some time. Actually, controllability implies that we can reach *any* given state starting from *any other* given state in *any* given interval of time. Thus given the states x_0, x_1 and the time interval t, let us find $u(\cdot)$ such that

$$x_1 = e^{At} x_0 + \int_0^t e^{A(t-s)} Bu(s) \, ds .$$

Let

$$z = x_1 - e^{At} x_0 .$$

Let us seek $u(\cdot)$ in the form

$$u(s) = B^* e^{A^*(t-s)} x , \qquad 0 < s < t ,$$

where we need to determine x. But $R(t)$ is nonsingular, and

$$z = R(t)x$$

has the unique solution

$$x = R(t)^{-1} z .$$

Let us next see what the controllability of the state space implies for the input-output description.

THEOREM 3.5. Suppose in the state-space description (3.1) of a system the state space is controllable. Then the system can be given the input-output representation

$$v(t) = \int_0^\infty W(\sigma) u(t-\sigma)\, d\sigma \qquad \text{for all } t , \tag{3.10}$$

where $W(\cdot)$ is the "weighting pattern" of the system.

PROOF. The main idea is that because of controllability, the state at the time a can be accounted for as due to some "previous" input. Thus in

$$\begin{aligned} v(t) &= Ce^{A(t-a)}x(a) + \int_a^t Ce^{A(t-\sigma)}Bu(\sigma)\, d\sigma + Du(t) \\ &= Ce^{A(t-a)}x(a) + \int_a^t W(t-\sigma)u(\sigma)\, d\sigma + Du(t) \end{aligned}$$

we note that we can write (because of controllability)

$$x(a) = \int_{a-L}^a e^{A(a-s)}Bu(s)\, ds$$

for any L and appropriate control $u(\cdot)$. Hence we can write

$$\begin{aligned} Ce^{A(t-a)}x(a) &= Ce^{A(t-a)} \int_{a-L}^a e^{A}(a-s)Bu(s)\, ds \\ &= \int_{a-L}^a W(t-s)u(s)\, ds . \end{aligned}$$

We can clearly replace $(a - L)$ by $-\infty$, by allowing $u(\cdot)$ to be zero for $-\infty < t < a - L$. Hence finally we get

$$v(t) = \int_{-\infty}^t W(t-s)\, ds + Du(t) = \int_0^\infty W(s)u(t-s)\, ds + Du(t) ,$$

as required, for all t. Note that to use representation (3.10) the entire "past history" of the input must be known. Also we may absorb D into $W(\cdot)$ by allowing δ-functions therein.

REMARK. It is possible to deduce a state-space representation, in which the state space is controllable, if we are given the input-output representation (3.10) (see [1]). We may say that a *system* is *controllable* if (3.10) holds, if we regard a system as being defined by the input-output relation.

DEFINITION. We shall say $(A \sim B)$ (where A is $n\times n$, B is $n\times p$) is "Controllable" if the Controllability matrix L_C:

$$| B \; AB \; \cdots \; A^{n-1} B |$$

has full rank.

Problem

Let T be a nonsingular $n\times n$ matrix. Show that if $(A \sim B)$ is controllable, so is

$$((TAT^{-1}) \sim TB) \; .$$

Also $(A \sim B)$ is controllable if and only if $((\lambda I - A) \sim B)$ is controllable for every λ.

SOLUTION:

$$\sum_0^{n-1} a_k A^k B = 0 \iff \sum_0^{n-1} a_k T A^k T^{-1} TB = 0$$

$$\iff \sum_0^{n-1} a_k (TAT^{-1})^k TB = 0 ,$$

$$\int_0^t e^{(\lambda I - A)s} BB^* e^{(\bar{\lambda} - A^*)s} x \, ds = 0$$

$$\iff \quad B^* e^{(\bar{\lambda} - A^*)s} x = 0 , \qquad 0 \le s \le t ,$$

$$\iff \quad B^* e^{-A^* s} x = 0 , \qquad 0 \le s \le t ,$$

$$\iff \quad (-A \sim B) \qquad \text{controllable} ,$$

$$\iff \quad (A \sim B) \qquad \text{controllable} .$$

Single Input-Single Output Systems

We now want to address the question: given an input-output description of a system can we tell whether it is possible to deduce an equivalent state-space description in which the state space is controllable? It turns out that we can indeed give a simple answer to this question for the case where both the input and output are *one*-dimensional.

THEOREM 3.6. Let a SI-SO system be described by the input-output relation†

$$\sum_0^n a_k \frac{d^k v(t)}{dt^k} = \sum_0^n b_k \frac{d^k u(t)}{dt^k} ,$$

where

$$a_n = 1 .$$

We can deduce an equivalent state-space description in which the state space is controllable if the "denominator" polynomial

$$D(\lambda) = \sum_0^n a_k \lambda^k$$

and the "numerator" polynomial

$$N(\lambda) = \sum_0^n b_k \lambda^k$$

have no common factors (have no roots in common).

PROOF. First of all we note that an equivalent state-space representation is, from Chapter 1,

$$\dot{x} = Ax + \tilde{B}u ,$$

$$v = Cx + Du ,$$

where

$$A = \begin{vmatrix} 0 & 1 & 0 & 0 & \cdots & 0 \\ 0 & 0 & 1 & 0 & \cdots & 0 \\ \cdots & & & & & \cdots \\ -a_0 & -a_1 & \cdots & & & -a_{n-1} \end{vmatrix} , \qquad \tilde{B} = \left[\sum_0^n b_k A^k B \right] ,$$

† Note the change from r to n.

where

$$B = \begin{vmatrix} 0 \\ 0 \\ \vdots \\ 0 \\ 1 \end{vmatrix} .$$

We shall show that this state space is Controllable. The controllability matrix is thus

$$[\tilde{B}, A\tilde{B}, ..., A^{n-1}\tilde{B}] .$$

Suppose the state space is not controllable; then

$$\sum_0^{n-1} c_k A^k \tilde{B} = 0$$

for some $\{c_k\}$, not all of which are zero. Let

$$p(\lambda) = \sum_0^{n-1} c_k \lambda^k .$$

Then

$$p(A)\tilde{B} = p(A)\left(\sum_0^n b_k A^k\right) B = 0 .$$

Suppose B is an eigenvector of the matrix

$$p(A)\left(\sum_0^n b_k A^k\right) .$$

Then it must be an eigenvector of A, which is impossible for the particular A and B we have constructed, as is readily verified. If B is not an eigenvector of

$$p(A)\left(\sum_0^n b_k A^k\right) ,$$

then the matrix

$$p(A)\left(\sum_0^n b_k A^k\right) = 0 .$$

Hence $\Delta(\lambda)$ must divide

$$p(\lambda)\left(\sum_0^n b_k \lambda^k\right) ,$$

or all the roots of $\Delta(\lambda)$ must be roots of

$$p(\lambda)\left[\sum_0^n b_k\lambda^k\right].$$

Since the minimal polynomial of A is also the characteristic polynomial of $\Delta(\lambda)$, we can find at least one root, say λ_1, such that

$$p(\lambda_1) \neq 0.$$

Hence

$$\sum_0^n b_k\lambda_1^k = 0,$$

or the numerator and denominator polynomials have a common root, or a (nontrivial) common factor—contradicting our hypothesis.

REMARK. The roots of $D(\lambda)$ are referred to as the "poles" of the system and the roots of $N(\lambda)$ are referred to as the "zeros" of the system. Also we say there is a "pole-zero cancellation" to mean that $D(\lambda)$ and $N(\lambda)$ have a common factor.

Let us show next that if $N(\lambda)$ and $D(\lambda)$ have a common root, then the state space we have constructed is not controllable. Let λ_0 denote the common root. We note that $D(\lambda)$ is the characteristic polynomial of A. Since $(\lambda - \lambda_0)$ is a factor of $D(\lambda)$, let $p(\lambda)$ denote the quotient polynomial

$$p(\lambda) = \frac{D(\lambda)}{\lambda - \lambda_0} = \sum_0^{n-1} c_k\lambda^k.$$

Now

$$p(\lambda)\left[\sum_0^n b_k\lambda^k\right] = D(\lambda)\left[\frac{\sum_0^n b_k\lambda^k}{\lambda - \lambda_0}\right].$$

Since $(\lambda - \lambda_0)$ also divides

$$\left[\sum_0^n b_k\lambda^k\right],$$

we see that $D(\lambda)$ divides the left side polynomial, and hence

$$p(A)\left[\sum_0^n b_kA^k\right] = 0.$$

Hence

$$\sum_{0}^{n-1} c_k A^k \tilde{B} = \left[\sum_{0}^{n-1} c_k A^k\right]\left[\sum_{0}^{n} b_k A^k\right] B = p(A)\left[\sum_{0}^{n} b_k A^k\right] B = 0 ,$$

or the state space is not Controllable.

Problems for Chapter 3

1. Suppose $(A \sim B)$ is Controllable. Show that for any λ

$$((\lambda I - A) \sim B)$$

is also Controllable.

2. Repeat Problem 4 of Chapter 1 for the system in Problem 5 of Chapter 1.

3. Determine a state-space representation for

$$\frac{d^3 v(t)}{dt^3} + \frac{dv}{dt} = \frac{du}{dt} + au(t) ,$$

where a is arbitrary. For what value of a is the system Controllable? Show this using your representation.

4. Determine whether the aircraft system in Chapter 1 is Controllable.

5. Consider a SI-SO system with state-space description (3.0). Show that it is not controllable for any B if the minimal polynomial of A is different from the characteristic polynomial.

6. Find necessary and sufficient conditions in order that $(A \sim A)$ is Controllable.

7. Specializing to the SI-SO case, show that the state space deducible from (2.26) is not Controllable if $n > r$.

8. Consider the state-space description (3.0) in which

$$A = \begin{vmatrix} 0 & 0 & 1 & 0 \\ 0 & 0 & 0 & 1 \\ -1 & 0 & -1 & 0 \\ 0 & -1 & 0 & -1 \end{vmatrix} , \qquad B = \begin{vmatrix} 0 & 0 \\ 0 & 0 \\ 1 & 0 \\ 0 & 1 \end{vmatrix} .$$

Is this Controllable?

9. Helicopter hover-control system (cf. [3]).

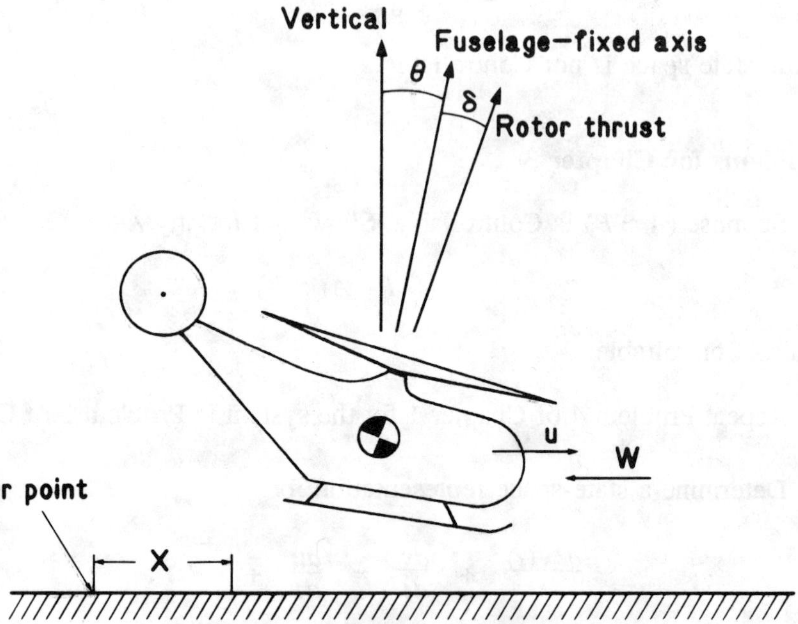

Nomenclature and reference axes for helicopter hover–control system

The figure shows the nomenclature and references axes. The four state variables are

θ = pitch angle of the fuselage,

q = pitch rate $(q = \dot{\theta})$,

x = horizontal distance of mass center from the hover points, and

u = horizontal velocity $(u = \dot{x})$.

The control variable is

θ = tilt angle of the rotor-thrust vector with respect to the fuselage.

The primary disturbance is horizontal wind, w. The vertical motions (controlled by increasing or decreasing rotor thrust), the lateral motions (controlled by tilting the rotor thrust laterally), and the yawing motions (controlled by the tail rotor) are all nearly uncoupled from the longitudinal equation of motion in state-variable form can be written as

$$\begin{vmatrix} \dot\theta \\ \dot q \\ \dot u \\ \dot x \end{vmatrix} = \begin{vmatrix} 0 & 1 & 0 & 0 \\ 0 & -\sigma_1 & -\alpha_1 & 0 \\ g & -\alpha_2 & -\sigma_2 & 0 \\ 0 & 0 & 1 & 0 \end{vmatrix} \begin{vmatrix} 0 \\ q \\ u \\ x \end{vmatrix} + \begin{vmatrix} 0 \\ n \\ g \\ 0 \end{vmatrix} \delta + \begin{vmatrix} 0 \\ -\alpha_1 \\ -\sigma_2 \\ 0 \end{vmatrix} w$$

$$\sigma_1 = .415\,, \quad \sigma_2 = .0198\,, \quad n = 6.27\,,$$

$$\alpha_1 = .0111\,, \quad \alpha_2 = 1.43\,, \quad g = 9.80\,.$$

Determine whether the tilt-angle control system is Controllable. Determine whether the wind-disturbance input system is Controllable.

References for Chapter 3

[1] Zadeh, L.A., and Polak, E. *System Theory*. New York: McGraw-Hill Inc., 1969.

[2] Wonham, W.M. *Linear Multivariable Control: A Geometric Approach*. New York Berlin Heidelberg: Springer-Verlag, 1979.

[3] Bryson, A.E., and Ho, Y.C. *Applied Optimal Control*. Blaisdell, 1969.

CHAPTER 4

LINEAR DYNAMIC SYSTEMS: OBSERVABILITY

Let a state-space description of a time-invariant linear dynamic system be given:

$$\left.\begin{aligned} \dot{x} &= Ax + Bu , \\ v &= Cx + Du \end{aligned}\right\} \tag{4.1}$$

(where A is $n \times n$, C is $m \times n$, to be specific). Let an input-output pair, ($u(t)$, $t \geq a$; $v(t)$, $t \geq a$) be given. Can we deduce from this what the state was at the initial time $t = a$?

Let us see what is involved in answering this question. Solving the input-output state equation, we obtain

$$x(t) = e^{A(t-a)}x(a) + \int_a^t e^{A(t-s)}Bu(s)\, ds \tag{4.2}$$

and hence

$$v(t) = Ce^{A(t-a)}x(a) + \int_a^t Ce^{A(t-s)}Bu(s)\, ds + Du(t) . \tag{4.3}$$

Since $u(\cdot)$ is given, we can subtract the terms containing the input and obtain

$$Ce^{A(t-a)}x(a) = v(t) - \int_a^t Ce^{A(t-s)}Bu(s)\, ds - Du(t) .$$

Note that the left side is the response to zero input in $t \geq a$. Thus we are required to find the state at time $t = a$, given the zero-input response in

$t \geq a$. Let us fix a nonzero time interval, $a \leq t \leq T + a$, say. Hence without loss of generality, our problem is to find the vector x given

$$v(t) = Ce^{At}x, \qquad 0 \leq t \leq T,$$

where $T > 0$.

Suppose now that there is a nonzero vector z such that

$$Ce^{At}z = 0, \qquad 0 \leq t \leq T.$$

Then

$$v(t) = Ce^{At}x = Ce^{At}(x + z), \qquad 0 \leq t \leq T,$$

and thus there can be more than one (in fact, infinitely many) initial state which matches the given response. In other words, the initial state is *not* "observable."

Observability

DEFINITION. The state space is said to be "observable" if for any $T > 0$

$$Ce^{As}x = 0, \qquad 0 \leq s \leq T,$$

implies that $x = 0$. We abbreviate this to $(C \sim A)$ observable.

THEOREM 4.1. The state space is observable if and only if

$$R(t) = \int_0^t e^{A^*s} C^* Ce^{As} \, ds \tag{4.4}$$

is nonsingular for some $t > 0$.

PROOF. Suppose the state space is observable. We shall show that $R(t)$ must be nonsingular for every $t > 0$. Suppose not. Suppose for some $t = t_0 > 0$:

$$R(t_0)x = 0, \qquad x \neq 0.$$

Then

$$\begin{aligned} 0 &= [R(t_0)x, x] \\ &= \int_0^{t_0} [e^{A^*s}C^*Ce^{As}x, \, x] \, ds \\ &= \int_0^{t_0} \|Ce^{As}x\|^2 \, ds. \end{aligned}$$

Hence

$$Ce^{As}x = 0, \qquad 0 \le s \le t_0,$$

which contradicts observability. Conversely, suppose $R(t)$ is nonsingular for some $t = t_1 > 0$. Suppose that the state space is *not* observable, so that for some $T > 0$

$$Ce^{As}x = 0, \qquad 0 \le s \le T, \ x \ne 0.$$

Then

$$\left.\frac{d^k(Ce^{As}x)}{ds^k}\right|_{s=0} = 0,$$

or

$$CA^k x = 0 \qquad \text{for every } k.$$

Hence

$$Ce^{As}x = 0 \qquad \text{for } every\ s, \ 0 \le s$$

(not only for $0 \le s \le T$!). Hence multiplying by $e^{A^*s}C^*$ on the left and integrating,

$$\int_0^t e^{A^*s}C^*Ce^{As}x\, ds = 0 = R(t)x$$

for *every* t, contradicting the hypothesis that $R(t_1)$ is nonsingular.

REMARK. We see that if $R(t)$ is nonsingular for some t, it is nonsingular for every t.

DEFINITION. The compound matrix $(mn \times n)$

$$\begin{vmatrix} C \\ CA \\ \vdots \\ CA^{n-1} \end{vmatrix}$$

is called the "observability" matrix.

THEOREM 4.2. The state space is observable if and only if the observability matrix has full rank.

PROOF. Suppose the observability matrix has full rank. Suppose the state space is not observable so that for some $t > 0$

$$Ce^{As}x = 0, \qquad 0 \le s \le t,\ x \ne 0 .$$

Then

$$\left. \frac{d^k (Ce^{As}x)}{ds^k} \right|_{s=0} = 0 = CA^k x .$$

Hence

$$0 = Cx = CAx = \cdots = CA^{n-1}x ,$$

or the observability matrix does not have full rank, which is a contradiction. Suppose the observability matrix does not have full rank so that

$$0 = Cx = CAx = \cdots = CA^{n-1}x$$

for some nonzero x. Then by the Cayley-Hamilton theorem

$$CA^n x = 0 \qquad \text{for every } n ;$$

hence

$$Ce^{As}x = 0 \qquad \text{for } 0 \le s \le T ,$$

for $T > 0$, which contradicts observability.

Finding the Initial State

Suppose now the state space is observable. Given the zero input response

$$v(s) = Ce^{As}x , \qquad 0 \le s \le t ,$$

we can find x for any t. Multiply on the left by $e^{A^*s}C^*$ and integrate. Then

$$\int_0^t e^{A^*s}C^* \, v(s)\, ds = R(t)x$$

and hence

$$x = R(t)^{-1} \int_0^t e^{A^*s}C^* v(s)\, ds .$$

Note that this holds for every t, however small.

Finally we note the "duality"

$$(C \sim A) \text{ observability } \leftrightarrow (A^* \sim C^*) \text{ controllability} .$$

Equivalence of Observable State Spaces

THEOREM 4.3. Suppose that a system described by an input-output relation has two equivalent (finite-dimensional) state-space descriptions

$$\left.\begin{aligned} \dot{x}(t) &= Ax(t) + Bu(t) , \\ v(t) &= Cx(t) + Du(t) \end{aligned}\right\} \quad (1)$$

and

$$\left.\begin{aligned} \dot{y}(t) &= Fy(t) + Gu(t) , \\ v(t) &= Hx(t) + \tilde{D}u(t) \end{aligned}\right\} \quad (2)$$

and furthermore, both state spaces are observable. Then the two state spaces are isomorphic; they have the same dimension and we can find a nonsingular matrix L such that

$$F = L^{-1}AL ,$$

$$H = CL , \qquad G = L^{-1}B ,$$

$$D = \tilde{D} .$$

PROOF. For any y in the state space (2),

$$v(s) = He^{Fs}y , \qquad s \geq 0 ,$$

defines a zero-input response. Hence there must be an element x in the state space corresponding to (1) such that

$$He^{Fs}y = Ce^{As}x , \qquad s \geq 0 . \tag{4.5}$$

There can only be one such element, because if there were two, say x_1, x_2, we would have

$$Ce^{Fs}x_1 = Ce^{As}x_2 = He^{Fs}y ,$$

and since the state space (1) is observable, x_1 must equal x_2. This defines a transformation or mapping defined on the state space (2) into that corresponding to (1):

$$x = Ly .$$

The mapping is quickly verified to be linear. Moreover, for any x in the state space (1), defining the zero-input response

$$Ce^{As}x, \qquad s \geq 0 ,$$

there must be a unique state y in state space (2) such that

$$Ce^{As}x = He^{Fs}y, \qquad s \geq 0 ,$$

and this implies that the two state spaces must be of the same dimension, say n, and hence that L can be represented by an $n \times n$ square matrix. Moreover, L is nonsingular since

$$Ly = 0$$

would imply that

$$He^{Fs}y = 0, \qquad s \geq 0 ,$$

and hence y must be zero, since y is arbitrary. Again for any $\Delta \geq 0$,

$$Ce^{At}(e^{A\Delta}x) = He^{Ft}(e^{F\Delta}y), \qquad t \geq 0 ,$$

it follows that

$$e^{A\Delta}x = Le^{F\Delta}y, \qquad \Delta \geq 0 ,$$

and taking the derivative at $\Delta = 0$, we get

$$Ax = LFy = LFL^{-1}x .$$

Let us now start with the zero state in each system and apply the same input $u(s)$ for $0 \leq s \leq \Delta$, and let $u(s) = 0$ for $s \geq \Delta$. Then

$$x(\Delta) = Ly(\Delta)$$

and hence

$$\int_0^{\Delta} e^{A(\Delta-\sigma)}Bu(\sigma)\, d\sigma = L\int_0^{\Delta} e^{F(\Delta-\sigma)}Gu(\sigma)\, d\sigma ;$$

and since Δ is arbitrary, we can differentiate both sides and obtain

$$Ax(\Delta) + Bu(\Delta) = LFy(\Delta) + LGu(\Delta) = LFL^{-1}x(\Delta) + LGu(\Delta) .$$

Since

$$A = LFL^{-1} ,$$

it follows that, since $u(\Delta)$ is arbitrary,

$$B = LG ,$$

and finally,

$$\tilde{D}u(t) + H\int_0^t e^{F(t-\sigma)}Gu(\sigma)\, d\sigma = \tilde{D}u(t) + CL\int_0^t e^{F(t-\sigma)}L^{-1}Bu(\sigma)\, d\sigma$$

$$= \tilde{D}u(t) + C\int_0^t e^{A(t-\sigma)}Bu(\sigma)\, d\sigma ,$$

which must equal

$$Du(t) + C\int_0^t e^{A(t-\sigma)}Bu(\sigma)\, d\sigma ,$$

or

$$\tilde{D} = D .$$

Finally, let us obtain a concrete representation for L, using (4.5)

Multiply on the left side by $e^{A^*s}C^*$ and integrate for some fixed $t > 0$:

$$\int_0^t e^{A^*s}C^*Ce^{As}x\, ds = \left[\int_0^t e^{A^*s}C^*He^{Fs}\, ds\right]y .$$

Or, letting

$$R(t) = \int_0^t e^{A^*s}C^*Ce^{As}\, ds ,$$

we have, since $R(t)$ is nonsingular,

$$x = \left[R(t)^{-1}\int_0^t e^{A^*s}C^*He^{Fs}\, ds\right]y .$$

Hence

$$L = R(t)^{-1} \int_0^t e^{A^*s} C^* H e^{Fs} \, ds .$$

We may also see directly that the right side is L by substituting in it

$$H = CL ,$$
$$F = L^{-1} A L .$$

Then

$$R(t)^{-1} \int_0^t e^{A^*s} C^* C L e^{L^{-1}ALs} \, ds = R(t)^{-1} \int_0^t e^{A^*s} C^* C L L^{-1} e^{As} L \, ds$$
$$= R(t)^{-1} R(t) L$$
$$= L .$$

As we have seen, given any input-output relation we can find more than one equivalent state-space description. We see that if the state space is observable, then there must be an isomorphism between them, as specified in Theorem 3.3.

DEFINITION. We shall say that two state-space descriptions are "input-output equivalent" if the input-output pairs are the same in each.

THEOREM 4.4. Given any state-space description of a time-invariant linear dynamic system, we can construct an input-output equivalent state-space description in which the state space is observable.†

PROOF. Let the given state-space description be

$$\left.\begin{aligned} \dot{x}(t) &= Ax(t) + Bu(t) , \\ v(t) &= Cx(t) \end{aligned}\right\} \tag{4.6}$$

(we omit the term $Du(t)$ since that plays no role here). Let

$$\Sigma_0 = [x \mid Ce^{At}x = 0, \, t \geq 0] . \tag{4.7}$$

† Such an observable state space is called "reduced," and the process is called "state reduction."

This is a subspace (of the state space). Let Σ_R denote its orthogonal complement. (Recall that the orthogonal complement of Σ_0 is the subspace of vectors orthogonal to Σ_0.) Let P denote the projection onto Σ_R. Note that for any state x,

$$x - Px \in \Sigma_0$$

and hence

$$Ce^{At}(x - Px) = 0, \qquad t \geq 0,$$

or

$$Ce^{At}x = Ce^{At}Px,$$

hence

$$CA^k x = CA^k Px, \qquad k \geq 0. \tag{4.8}$$

In particular,

$$Cx = CPx. \tag{4.9}$$

Next suppose

$$x \in \Sigma_0$$

so that

$$Ce^{At}x = 0, \qquad t \geq 0.$$

Then also for any $\Delta > 0$

$$0 = Ce^{A(s+\Delta)}x = Ce^{As}e^{A\Delta}x, \qquad s \geq 0.$$

Hence it follows that

$$e^{At}\Sigma_0 \subset \Sigma_0 \qquad \text{for every } t.$$

Hence for every x

$$Pe^{At}(x - Px) = 0,$$

or

$$Pe^{At}x = Pe^{At}Px.$$

Differentiating and setting $t = 0$, we get

$$PA^k x = PA^k Px, \qquad k \geq 0 . \tag{4.10}$$

Now by (4.9)

$$v(t) = Cx(t) = CPx(t)$$

and from (4.6)

$$\begin{aligned} P\dot{x}(t) &= PAx(t) + PBu(t) \\ &= PAPx(t) + PBu(t) , \qquad \text{using (4.10)} . \end{aligned}$$

Hence letting

$$y(t) = Px(t) ,$$

we have

$$\left.\begin{aligned} \dot{y}(t) &= (PAP)y(t) + (PB)u(t) , \\ v(t) &= (CP)y(t) . \end{aligned}\right\} \tag{4.11}$$

But taking $u(t) = 0$, we have

$$\begin{aligned} y(t) &= Pe^{At}x(0) , \\ \dot{y}(t) &= PAy(t) , \\ &= PAPy(t) ; \end{aligned}$$

and hence

$$y(t) = e^{(PAP)t}Px(0) ,$$

or

$$Pe^{At}x(0) = e^{(PAP)t}Px(0) ,$$

or

$$Pe^{At}x = e^{(PAP)t}Px \qquad \text{for every } x \text{ and } t .$$

Hence also

$$Pe^{As}B = e^{(PAP)s}PB .$$

Moreover, we have in (4.11) a new state-space description with PAP in place of A, PB in place of B, and CP in place of C, and the new state space is Σ_R, the range of P. Every input-output pair of (4.6) is preserved in (4.11) by the very manner of our construction. Conversely, given any input-output pair given by (4.11), we have

$$\begin{aligned} v(t) &= CPe^{(PAP)t}y(0) + C\int_0^t e^{(PAP)(t-s)}PBu(s)\,ds \\ &= CPe^{(PAP)t}y(0) + CP\int_0^t e^{A(t-s)}Bu(s)\,ds , \end{aligned}$$

(and since we can write $y(0) = Px$)

$$\begin{aligned} &= CP\left\{e^{At}x + \int_0^t e^{A(t-s)}Bu(s)\,ds\right\} \\ &= C\left\{e^{At}x + \int_0^t e^{A(t-s)}Bu(s)\,ds\right\} \\ &= Cx(t) , \end{aligned}$$

where

$$\dot{x}(t) = Ax(t) + Bu(t) .$$

Hence every input-output pair of (4.11) is also an input-output pair of (4.6). The two state-space descriptions are thus input-output equivalent. It is immediate that the new state space is observable. For

$$0 = CPe^{(PAP)t}y = CPe^{(PAP)t}Px = Ce^{At}x , \qquad t \geq 0 ,$$

implies $x \in \Sigma_0$ and hence

$$y = Px = 0 .$$

Note that the Σ_0 is the null space of the observability matrix

$$L_o = \begin{vmatrix} C \\ CA \\ \vdots \\ CA^{n-1} \end{vmatrix}$$

mapping R^n into R^{nm}. Hence Σ_R, the orthogonal complement, is the range space of L_o^*, or equivalently, is the range of the $n \times n$ matrix

$$L_o^* L_o = \sum_0^{n-1} A^{*k} C^* C A^k .$$

In particular, Σ_0 is the null space of this matrix. Also

$$\Sigma_R = \text{range of} \int_0^t e^{A^*s} C^* C e^{As} \, ds \qquad \text{for any } t > 0 ,$$

since

$$\text{range of } R(t) = \text{range of } L_o^* = \text{range of } L_o^* L_o .$$

Let us next see how to characterize P, or the reduced state-space representation (4.11), in terms of C and A. First we have

LEMMA 4.1. We can find an integer $q \leq n$ such that the matrices C, CA, ..., CA^{q-1} are linearly independent while C, CA, ..., CA^{q-1}, CA^{q+k} are linearly dependent for $k \geq 0$.

PROOF. Let $\Delta(\lambda)$ be the characteristic polynomial of A

$$\Delta(\lambda) = \lambda^n + b_{n-1}\lambda^{n-1} + \cdots + b_0 .$$

Then

$$CA^n = \sum_0^{n-1} (-b_j) CA^j ,$$

hence C, CA, ..., CA^n are linearly dependent. Suppose q is the smallest (or the "first") integer such that C, CA, ..., CA^{q-1}, CA^q are linearly dependent. Then of course C, CA, ..., CA^{q-1} must be linearly independent and

$$CA^q = \sum_0^{q-1} a_j CA^j , \tag{4.12}$$

or equivalently

$$A^{*q}C^* = \sum_0^{q-1} \bar{a}_j A^{*j} C^* .$$

Hence multiplying on both sides by A^*, we have

$$A^{*q+1} C^* = \sum_0^{q-1} \bar{a}_j A^{*j+1} C^* ,$$

or

$$\begin{aligned} CA^{q+1} &= \sum_0^{q-1} a_j CA^{j+1} , \\ &= \sum_1^{q-1} a_{j-1} CA^j + a_{q-1} CA^q \\ &= \sum_1^{q-1} a_{j-1} CA^j + a_{q-1} \sum_0^{q-1} a_j CA^j \\ &= \sum_0^{q-1} c_j CA^j \end{aligned}$$

for appropriate c_j. Hence CA^{q+1} can be expressed in terms of C, CA, ..., CA^{q-1}. For higher powers, CA^{q+k}, $k > 1$, we can clearly proceed by induction.

COROLLARY 1. Suppose

$$CA^j x = 0 , \qquad j = 0, ..., q-1 .$$

Then $x \in \Sigma_0$.

PROOF. Follows from (4.12).

COROLLARY 2. Let v_0, ..., v_q be arbitrary elements in R^m. Then

$$CA^j x = v_j , \qquad j = 0, ..., q-1 , \tag{4.13}$$

has *at most* one solution in Σ_R.

PROOF. If (4.13) has more than one solution, their difference would be in Σ_0 by Corollary 1.

But it should be noted that (4.13) need *not* have a solution in general.

For a counterexample we only need to take A such that the minimal polynomial of A is

$$(\lambda^2 + \lambda)$$

and

$$C = I .$$

In this case $(C \sim A)$ is observable, and $q = 2$. But

$$Cx = x = v_0 ,$$

$$CAx = Ax = v_1 ,$$

has no solution for *arbitrary* v_0 and v_1; we have only to choose

$$Av_0 \neq v_1 ;$$

and this is possible since A is singular. For example, let v_0 be such that

$$Av_0 = 0$$

and v_1 is nonzero. Note that the dimension of Σ_R in this example is n while q is 2.

However, we do have:

THEOREM 4.5. Suppose we specialize to a single output system ($m = 1$, C is $n\times 1$). Then the rank of the observability matrix is equal to q and

$$CA^j x = v_j , \qquad j = 0, \ldots, q-1 , \tag{4.14}$$

for arbitrary v_j in R^1 *has* a unique solution in Σ_R.

PROOF. Every element of Σ_R can be expressed

$$\sum_0^{n-1} v_j A^{*j} C^* ,$$

where $v_j \in R^1$. The dimension of Σ_R is thus the number of linearly independent vectors among C^*, $A^* C^*$, ..., $A^{*n-1} C^*$, and hence is equal to q. Moreover, every element Σ_R can be expressed

$$\sum_0^{q-1} v_j A^{*j} C^* .$$

Let it be required to solve

$$CA^j x = a_j , \qquad j = 0,\dots,q-1 . \tag{4.15}$$

For x in Σ_R, we have then

$$CA^k \sum_0^{q-1} v_j A^{*j} C^* = a_k , \qquad k = 0, \dots, q-1 ,$$

or

$$\sum_0^{q-1} [A^{*k} C^*, A^{*j} C^*] v_j = a_k , \qquad k = 0, \dots, q-1 ;$$

but the matrix R with components

$$r_{kj} = \{[A^{*k} C^*, \; A^{*j} C^*]\} , \qquad k = 0, \dots, q-1 ,$$

is nonsingular, being the moment matrix of the linearly independent vectors $C^*, A^* C^*, \dots, A^{*q-1} C^*$. Hence (4.14) has a unique solution in Σ_R for arbitrary $\{a_j\}$. The dimension of the range of

$$\sum_0^{q-1} A^{*j} C^* C A^j$$

is q, and the same as that of

$$\sum_0^{n-1} A^{*j} C^* C A^j .$$

State-Space Reduction: Single Output Case

Using (4.14) we can obtain a canonical form for the observable state space of a single output system. Thus let us define a mapping on R^n onto R^q defined by

$$Ex = z ; \qquad CA^j x = v_j , \qquad j = 0, 1, \dots, q-1 ,$$

$$z = \begin{vmatrix} v_0 \\ \vdots \\ v_{q-1} \end{vmatrix} .$$

Note that E is 1:1 on $\Sigma_R \to R^q$. It can thus be represented by a nonsingular $q \times q$ matrix depending on the choice of our basis for Σ_R. The representation would be the matrix R, if we choose $\{A^{*j} C^*\}$ as the basis. (Of course E is not a projection, since the range space is different from R^n.)

Let $x \in \Sigma_R$ and

$$Ex = z = \begin{vmatrix} v_0 \\ \vdots \\ v_{q-1} \end{vmatrix} .$$

Let us determine $\mathcal{A}$ such that

$$EAx = \mathcal{A}z = \begin{vmatrix} b_0 \\ \vdots \\ b_{q-1} \end{vmatrix} .$$

This means that

$$CA^jAx = b_j , \qquad j = 0, 1, \ldots, q-1 ,$$

or

$$\begin{aligned} b_j &= v_{j+1} , & j &= 0, \ldots, q-2 , \\ &= CA^q x , & j &= q-1 . \end{aligned}$$

Let

$$CA^q = \sum_0^{q-1} a_j CA^j .$$

Then

$$b_{q-1} = \sum_0^{q-1} a_j v_j ;$$

hence

$$\mathcal{A} = \begin{vmatrix} 0 & 1 & 0 & \cdots & 0 \\ 0 & 0 & 1 & \cdots & 0 \\ \vdots & \vdots & \vdots & & \vdots \\ a_0 & a_1 & a_2 & \cdots & a_{q-1} \end{vmatrix} . \tag{4.16}$$

Next,

$$v = Cx = (CE^{-1})z = v_0 ,$$

hence

$$\tilde{C} = CE^{-1} = \text{row}\,[1, 0, ..., 0] .$$

Finally

$$E(Bu) = z ,$$

or

$$CA^j Bu = v_j , \qquad j = 0, ..., q-1 .$$

Let us represent this as

$$\tilde{B}u = z .$$

Hence we have the observable state-space representation

$$\left.\begin{aligned} \dot{z} &= \mathcal{A}z + \tilde{B}u , \\ v &= \tilde{C}z ; \end{aligned}\right\} \tag{4.17}$$

$$\mathcal{A} = \begin{vmatrix} 0 & 1 & 0 & \cdots & 0 \\ 0 & 0 & 1 & \cdots & 0 \\ \vdots & \vdots & \vdots & & \vdots \\ a_0 & a_1 & a_2 & \cdots & a_{q-1} \end{vmatrix} ; \qquad C = \text{row}\,[1, 0, ..., 0] .$$

Let us prove the input-output equivalence to the original state space. Suppose we begin with (4.17). Let

$$x(t) = E^{-1} z(t) ,$$

where $x(t) \in \Sigma_R$. Then

$$\dot{x}(t) = E^{-1}\mathcal{A}Ex(t) + E^{-1}\tilde{B}u(t) .$$

With P denoting the projection corresponding to Σ_R, we have

$$\dot{x}(t) = P(Ax(t) + Bu(t)) .$$

Since

$$\dot{x}(t) \in \Sigma_R ,$$

we have

$$P(\dot{x}(t) - Ax(t) - Bu(t)) = 0 ,$$

or

$$\dot{x}(t) - Ax(t) - Bu(t) = h(t) ,$$

where

$$h(t) \in \Sigma_0 .$$

Let us solve

$$\dot{y}(t) - Ay(t) = h(t) ,$$

$$y(0) = 0 .$$

Then

$$y(t) = \int_0^t e^{A(t-\sigma)} h(\sigma)\, d\sigma \in \Sigma_0 .$$

Hence

$$\frac{d}{dt}(x(t) - y(t)) = A(x(t) - y(t)) + Bu(t) ,$$

$$\mathrm{v}(t) = \tilde{C}z(t) = \tilde{C}Ex(t) = CPx(t) = C(x(t) - y(t)) .$$

Hence any $(u(\cdot), \mathrm{v}(\cdot))$ pair yielded by (4.17) can be obtained in the original state-space representation.

Conversely,

$$z(t) = Ex(t) ,$$

where $x(t)$ satisfies (4.6), will of course satisfy (4.17).

The representation (4.17) will be called the "canonical representation" for a single output system.

We note that corresponding to (4.17) we also have the input-output relation

$$\frac{d\mathrm{v}^q(t)}{dt^q} + \sum_0^{q-1} a_j \frac{d\mathrm{v}^j(t)}{dt^j} = \sum_0^q b_j \frac{du^j(t)}{dt^j} , \tag{4.18}$$

where the b_j are $1\times p$ matrices, the input $u(\cdot)$ being $p\times 1$. The equivalence of this to (4.17) can be shown in the same way as we did in going from (4.17) to (4.6).

EXAMPLE. Let the state-space representation be

$$\left.\begin{aligned} \dot{x}(t) &= Ax(t) , \\ \mathrm{v}(t) &= x(t) , \end{aligned}\right\} \tag{4.19}$$

where A is $n\times n$ (B is zero and $C = I$). Let the minimal polynomial of A be

$$\phi(\lambda) = \lambda^2 + \lambda .$$

Then, since $C = I$,

$$CA^2 = -CA \tag{4.20}$$

or q in (4.12) is 2. Since

$$\mathrm{v}(t) = x(t) ,$$

$$\dot{\mathrm{v}}(t) = Ax(t) ,$$

$$\ddot{\mathrm{v}}(t) = A^2 x(t) ,$$

we obtain the input-output relation

$$\ddot{\mathrm{v}}(t) + \dot{\mathrm{v}}(t) = 0 , \tag{4.21}$$

using (4.20).

However (4.21) is not equivalent to (4.19), since (4.21) has the solution

$$\mathrm{v}(t) = (\mathrm{v}(0) + \dot{\mathrm{v}}(0)) - e^{-t}\,\dot{\mathrm{v}}(0) , \tag{4.22}$$

while (4.19) has the solution

$$\mathrm{v}(t) = x(t) = e^{At}\mathrm{v}(0) . \tag{4.23}$$

Hence the input-output pairs satisfying (4.22), with $\dot{\mathrm{v}}(0)$ *not* equal to $A\mathrm{v}(0)$, cannot satisfy (4.19) and hence would have an observable state-space dimension of $(2n)$, while that of (4.19) is only n.

State Reduction: Multi-Output Case

Let us now describe the state-reduction procedure in the general case $m \neq 1$. We note that Σ_0 is orthogonal to all the rows in the observability matrix L_0. Hence Σ_R is spanned by the linearly independent rows in

$$\begin{vmatrix} C \\ CA \\ \vdots \\ CA^{p-1} \end{vmatrix} .$$

If the dimension of Σ_R is r, then there must be r linearly independent rows. Let $e_1, ..., e_r$ be a basis for the columns in

$$| C^* \ \cdots \ A^{*p-1} C^* | .$$

Then with P denoting the projection corresponding to Σ_R, we have that

$$Px = \sum_1^r a_k e_k ,$$

where the coefficients $\{a_k\}$ are determined by

$$[x, e_j] = \sum_1^r a_k [e_k, e_j] , \qquad j = 1,...,r .$$

Hence

$$PAx = \sum_1^r v_k e_k ,$$

where the $\{v_k\}$ are determined from

$$PAx = PAPx = P \sum_1^r a_k A e_k .$$

Hence

$$\sum_1^r a_k [Ae_k, e_j] = \sum_1^r v_k [e_k, e_j] , \quad j = 1,...,r ,$$

and if

$$M = \{[Ae_k, e_j]\} ,$$

$$R = \{[e_k, e_j]\} ,$$

$$a = \left| \begin{matrix} a_1 \\ \vdots \\ a_r \end{matrix} \right| , \qquad y = \left| \begin{matrix} v_1 \\ \vdots \\ v_r \end{matrix} \right| ,$$

we have

$$y = R^{-1} Ma .$$

Next,

$$Cx = CPx = \sum_1^r a_k C e_k = \tilde{C} a ,$$

where

$$\tilde{C} = | \, Ce_1 \; Ce_2 \cdots Ce_r \, | .$$

Finally, we can write

$$PBu = \tilde{B}u ,$$

where

$$\tilde{B}u = R^{-1} \left| \begin{matrix} e_1^* \\ \vdots \\ e_r^* \end{matrix} \right| Bu .$$

Hence our reduced state-space equations are

$$\left. \begin{aligned} \dot{z} &= (R^{-1}M) z(t) + \tilde{B}u(t) , \\ v(t) &= \tilde{C} z(t) , \end{aligned} \right\} \tag{4.24}$$

where $z(\cdot)$ is $r \times 1$; and (4.24) is input-output equivalent to (4.6).

Let us next connect Observability and Controllability.

THEOREM 4.6. Suppose a linear dynamic system has a state-space description in which the state space is controllable. Then we can obtain an input-output equivalent state-space description in which the state space is both observable and controllable.

PROOF. Let Σ denote the given state space which is controllable. By Theorem 3.4 we can find a subspace Σ_R of Σ which is observable and input-output equivalent. Since Σ_R is a subspace, it is clearly controllable, too.

We are now ready to prove the converse of Theorem 3.6.

THEOREM 4.7. Let a SI-SO system have the input-output relation

$$\sum_0^n a_k \frac{d^k v(t)}{dt^k} = \sum_0^n b_k \frac{d^k u(t)}{dt^k} , \tag{4.25}$$

where

$$a_n = 1 .$$

Let

$$\begin{aligned} \dot{x}(t) &= Fx(t) + Gu(t) , \\ v(t) &= Hx(t) + Du(t) \end{aligned} \tag{4.26}$$

be an equivalent state-space representation in which the state space is observable. Then the observable state space is controllable if and only if there is no pole-zero cancellation.

PROOF. Let us recall the state-space representation of (2.14). Let us verify that this state space is observable. This is immediate from the fact that

$$\begin{aligned} C &= 1 \quad 0 \quad 0 \quad 0 \quad \cdots \quad 0 \\ CA &= 0 \quad 1 \quad 0 \quad 0 \quad \cdots \quad 0 \\ \vdots & \quad\ \vdots \quad\ \vdots \quad\ \vdots \quad\ \vdots \qquad\ \ \vdots \\ CA^{n-1} &= 0 \quad 0 \quad 0 \quad 0 \quad \cdots \quad 1 \end{aligned}$$

and hence are linearly independent. By Theorem 3.6, this state space is controllable. Now by Theorem 4.3 there is a 1:1 isomorphism between state spaces which are observable and input-output equivalent, and further for some nonsingular matrix T:

$$A = TFT^{-1} ,$$

$$\tilde{B} = TG ;$$

but

$$(A \sim \tilde{B}) \text{ is controllable}$$

if and only if

$$(T^{-1}AT \sim T^{-1}B)$$

or

$$(F \sim G) \text{ is controllable .}$$

Hence it follows that $(F \sim G)$ is controllable if and only if there is no pole-zero cancellation.

Problems for Chapter 4

1. Suppose $(C \sim A)$ is observable. Show that $(C \sim (\lambda I - A))$ is also observable for any λ; and conversely.

2. In the aircraft problem suppose the output is restricted to be only the pitch rate $d\theta/dt$. Is the state still observable? Which of the two outputs, pitch, pitch-rate, angle-of-attack, would you pick and still maintain state observability?

3.

$$A = \begin{vmatrix} -2 & 0 & 3 & -2 \\ 1 & -0.5 & 4 & 7 \\ 4 & 0 & 5 & 4 \\ 0 & 0 & 0 & -1 \end{vmatrix} ; \qquad C = \begin{vmatrix} 1 & -1 & 2 & 0 \\ 0 & 1 & -1 & 2 \\ 2 & -1 & 0 & 3 \end{vmatrix} .$$

Find the rank of the observability matrix. Find the smallest p such that

$$CA^p = \sum_0^{p-1} a_j CA^j .$$

Characterize the unobservable subspace, and characterize the observable (reduced state) space.

4. Show that the state space (2.14) in the multi-input-multi-output case is observable. ($| C \; CA \cdots CA^{r-1} |$ already has full rank.)

5. Consider a SI-SO system with state-space description as in (4.1). Show that the state space is not observable if the minimal polynomial of A is different from the characteristic polynomial, whatever C may be.

6. Consider a single input system with multi-output (the output v(·) is $m\times 1$ where $m \geq 2$) and input-output relation (4.25). Show that the (observable!) state space (2.14) (and hence any equivalent state space) cannot be controllable.

Hint: In the notation of (2.14) of Chapter 2,

$$\psi(A)\tilde{B} = 0 \,, \quad \tilde{B} \text{ is } mn\times 1 \,,$$

and the characteristic polynomial is $\psi(\lambda)^m$.

CHAPTER 5

LINEAR DYNAMIC SYSTEMS: STABILITY

Historically, the notion of Stability predates the notion of Controllability or Observability, the classical treatise of Liapunov, *Stability of Motion* (1892), having already laid the foundations of the theory.

Stability

DEFINITION. Given a state-space description of a time invariant linear dynamic system, we shall say that the state space is "stable" (in the sense of Liapunov) if the state of the system for zero input tends to zero asymptotically.

In other words, given the state-space equation, for zero input

$$\dot{x}(t) = Ax(t), \qquad t \geq 0,$$

"stability" means that the solution $x(t)$ is such that

$$\lim_{t\to\infty} \|x(t)\| = 0,$$

whatever the initial state.

Another way of saying this is that the system (now including the given state-space description) returns to rest† eventually, left to itself (that is, without "external stimulus").

We have already seen that:

THEOREM 5.1. Stability is equivalent to the real parts of all eigenvalues of A being (strictly) less than zero.

---

† We sometimes refer to the zero state as the "rest" state since the system remains at that state with zero input.

REMARK. Sometimes we call a system "marginally stable" if some eigenvalues have zero real parts and the remaining are negative. In other words, "stability" allows us to inlude eigenvalues whose real parts are merely nonpositive. We shall *not* do so here, since for our purposes we do need the stronger notion.

We shall also contract "state space is stable" to "A is stable," since only A is involved in the argument.

Next we need a basic estimate for the rate of growth of the norm of the solution of the state equation with zero input

$$\dot{x}(t) = Ax(t) .$$

For this purpose we note that

$$\begin{aligned} \frac{d}{dt}\|x(t)\|^2 &= \frac{d}{dt}[x(t), x(t)] \\ &= [Ax(t), x(t)] + [x(t), Ax(t)] \\ &= [x(t), (A+A^*)x(t)] . \end{aligned}$$

Since $(A+A^*)$ is self-adjoint, its eigenvalues are real. Moreover, if

$$\mu = \text{largest eigenvalue of } (A+A^*) ,$$

$$\lambda = \text{smallest eigenvalue of } (A+A^*) ,$$

then

$$\lambda[x, x] \le [(A+A^*)x, x] \le \mu[x, x] ,$$

hence

$$\lambda \|x(t)\|^2 \le \frac{d}{dt}\|x(t)\|^2 \le \mu \|x(t)\|^2 . \tag{5.1}$$

Let

$$\|x(t)\|^2 = m(t)$$

so that

$$\|x(0)\|^2 = m(0) ;$$

and we have first from (5.1) that

$$\frac{d}{dt}m(t) \leq \mu m(t) ,$$

or

$$\frac{d}{dt}m(t) - \mu m(t) \leq 0 .$$

Hence multiplying by the "integrating factor" $e^{-\mu t}$ we have

$$e^{-\mu t}\frac{d}{dt}m(t) - \mu e^{-\mu t} m(t) \leq 0 ,$$

or

$$e^{-\mu t}\frac{d}{dt}m(t) + \left(\frac{d}{dt}e^{-\mu t}\right) m(t) \leq 0 ,$$

or

$$\frac{d}{dt}[e^{-\mu t}m(t)] \leq 0 .$$

Hence integrating both sides from zero to t:

$$e^{-\mu t}m(t) - m(0) \leq 0 .$$

By parallel arguments

$$\frac{d}{dt}m(t) \geq \lambda m(t)$$

in (5.1) leads to

$$e^{-\lambda t}m(t) - m(0) \geq 0 .$$

Or, combining the inequalities:

$$m(0)e^{\lambda t} \leq \|x(t)\|^2 \leq m(0)e^{\mu t} ,$$

or

$$\|x(0)\|e^{(\lambda/2)t} \leq \|x(t)\| \leq \|x(0)\|e^{(\mu/2)t} . \qquad (5.2)$$

This is an important inequality for us, and in particular we can derive a sufficient condition for stability therefrom.

THEOREM 5.2. The state space with system matrix A is stable if the eigenvalues of $(A+A^*)$ are all less than zero.

The condition is not necessary, since we can readily construct a stable matrix A such that $(A+A^*)$ is not stable. For example, let

$$A = \begin{vmatrix} 0 & 1 \\ -3 & -4 \end{vmatrix}.$$

Then

$$\Delta(\lambda) = \lambda^2 + 4\lambda + 3$$

and A has eigenvalues -1 and -3. But

$$A + A^* = \begin{vmatrix} 0 & -2 \\ -2 & -8 \end{vmatrix}$$

and has one eigenvalue: $(\sqrt{20} - 4)$ which is positive. However, we can state:

THEOREM 5.3. Suppose A is stable. Then we can find a covariance P such that

$$AP + PA^*$$

is stable.

PROOF. Since A is stable, we can define

$$P = \int_0^\infty e^{As} e^{A^*s} \, ds ,$$

and we can readily verify that

$$AP = \int_0^\infty \left(\frac{d}{ds} e^{As} \right) e^{A^*s} \, ds ,$$

and integrating by parts,

$$= -I - \int_0^\infty e^{As} \frac{d}{ds} e^{A^*s} \, ds$$

$$= -I - PA^*$$

$$AP + PA^* = -I ,$$

which is clearly stable. We can also generalize Theorem 5.2.

THEOREM 5.4. Suppose for some covariance P,

$$AP + PA^*$$

is stable. Then A is stable.

PROOF. Let

$$AP + PA^* = -Q .$$

Then Q is self-adjoint, and because $(-Q)$ is given to be stable, Q is clearly a nonsingular covariance. Let λ be eigenvalue of A^*

$$A^* x = \lambda x , \qquad x \neq 0 .$$

Then

$$0 = [APx, x] + [PA^* x, x] + [Qx, x] = 2(\text{Re } \lambda)[Px, x] + [Qx, x] .$$

Hence, if

$$\text{Re } \lambda \geq 0 ,$$

we must have

$$[Qx, x] = 0$$

or

$$Qx = 0 ,$$

which is impossible.

Asymptotic Bound

The left side of (5.2) is not of much use in establishing stability, as we have seen. We can however obtain an asymptotic version, which is useful. Let

$$\omega_0 = \text{Max Re (spectrum of } A) .$$

Then we have:

THEOREM 5.5. Given any $\varepsilon > 0$, we can find M_ε such that

$$\| e^{At} \|_o \leq M_\varepsilon \, e^{(\omega_0 + \varepsilon)t} . \tag{5.3}$$

PROOF. First of all we recall that

$$\lim_{n\to\infty} \|e^{nAt}\|_o^{1/n} = \text{spectral radius of } e^{At} = e^{\omega_0 t} . \tag{5.4}$$

Let

$$m(t) = \log \|e^{At}\|_o .$$

Then we shall show that

$$\lim_{t\to\infty} \frac{m(t)}{t} = \omega_0 .$$

For this purpose, let us note that

$$\|e^{At}\|_o \geq e^{\omega_0 t}$$

and hence

$$\inf_{t>0} \frac{m(t)}{t} = \alpha > -\infty .$$

Moreover, given any $\varepsilon > 0$, we can find $t_\varepsilon > 0$ such that

$$\frac{m(t_\varepsilon)}{t_\varepsilon} = \alpha + \varepsilon .$$

Now, because

$$\|e^{A(t_1+t_2)}\|_o \leq \|e^{At_1}\|_o \|e^{At_2}\|_o ,$$

we can see that

$$m(t_1 + t_2) \leq m(t_1) + m(t_2) . \tag{5.5}$$

Further, any t has the "modulo t_ε" representation

$$t = nt_\varepsilon + \beta , \qquad 0 \leq \beta < t_\varepsilon .$$

Hence we have

$$\frac{m(t)}{t} = \frac{m(nt_\varepsilon + \beta)}{nt_\varepsilon + \beta} \leq \frac{m(nt_\varepsilon)}{nt_\varepsilon + \beta} + \frac{m(\beta)}{nt_\varepsilon + \beta}$$

$$\leq \frac{m(nt_\varepsilon)}{nt_\varepsilon} + \frac{m(\beta)}{nt_\varepsilon + \beta} ,$$

which, using (5.5),

$$\leq \frac{m(t_\varepsilon)}{n} + \frac{m(\beta)}{nt_\varepsilon + \beta} .$$

Clearly $m(t)$ is bounded in $[0, t_\varepsilon]$. Hence we see that

$$\lim_{t\to\infty} \frac{m(t)}{t} \leq \frac{m(t_\varepsilon)}{t_\varepsilon} \leq \alpha + \varepsilon ,$$

where ε is arbitrary. Hence

$$\lim_{t\to\infty} \frac{m(t)}{t} \leq \alpha .$$

But by the definition of α,

$$\frac{m(t)}{t} \geq \alpha$$

and hence

$$\lim_{t\to\infty} \frac{m(t)}{t} = \alpha .$$

Now from (5.4)

$$\lim_{n\to\infty} \frac{m(nt)}{nt} = \omega_0$$

and hence

$$\lim_{t\to\infty} \frac{m(t)}{t} = \alpha = \omega_0 .$$

Hence given $\varepsilon > 0$, we can find T_ε such that

$$\frac{m(t)}{t} - \omega_0 \leq \varepsilon \qquad \text{for all } t > T_\varepsilon ,$$

or

$$\frac{m(t)}{t} \leq \omega_0 + \varepsilon \qquad \text{for all } t > T_\varepsilon ,$$

or

$$m(t) \leq t(\omega_0 + \varepsilon) \qquad \text{for all } t > T_\varepsilon .$$

But since $m(t)$ is bounded in $[0, T_\varepsilon]$, we can find $M_\varepsilon > 1$ such that

$$m(t) \le \log M_\varepsilon + t(\omega_0 + \varepsilon) , \qquad 0 \le t \le T_\varepsilon .$$

Hence for all $t > 0$

$$m(t) \le \log M_\varepsilon + t(\omega_0 + \varepsilon)$$

or

$$\|e^{At}\|_o \le M_\varepsilon e^{t(\omega_0+\varepsilon)} .$$

COROLLARY. If $\omega_0 = -\sigma < 0$, for any γ, $0 < \gamma < \sigma$, we can find M_γ such that

$$\|e^{At}x\| \le M_\gamma e^{-\gamma t} \|x\| \qquad \text{for all } t > 0 .$$

PROOF. We have only to choose ε in the Theorem such that

$$-\sigma + \varepsilon = -\gamma , \qquad \gamma > 0 .$$

REMARK 1. Note that in Theorem 5.5 we have an alternate proof of Theorem 5.1 which does not specifically use the spectral representation (1.5) for e^{At}.

REMARK 2. Suppose A is stable and $(A+A^*)$ is not. Then we can find a state x such that

$$[(A+A^*)x,\ x] > 0 .$$

Hence

$$\frac{d}{dt}[e^{At}x,\ e^{At}x]\Big|_{t=0} = [(A+A^*)x,\ x] > 0 .$$

Hence we can find t_0 small enough so that

$$\|e^{At_0}x\|^2 > \|x\|^2 .$$

In fact, we can state:

THEOREM 5.6. A necessary and sufficient condition that

$$\|e^{At}\|_o \leq 1$$

for all t is that

$$-(A + A^*)$$

is a covariance.

PROOF. If $-(A+A^*)$ is a covariance, then

$$[(A+A^*)x,\ x] \leq 0$$

for every x, and hence, for all $t \geq 0$

$$\frac{d}{dt}[e^{At}x,\ e^{At}x] \leq 0$$

so that

$$\|e^{At}x\| \leq \|x\|$$

or

$$\|e^{At}\|_o \leq 1\ .$$

Conversely, if $\|e^{At}\|_o \leq 1$, then

$$[e^{At}x,\ e^{At}x] \leq [x, x]$$

for every x, and hence

$$\frac{d}{dt}[e^{At}x,\ e^{At}x]\Big|_{t=0} \leq 0$$

or

$$[(A+A^*)x,\ x] \leq 0\ .$$

Stable States

Even though a system may not be stable, there may be states which are stable. Thus:

DEFINITION. A state x is said to be stable if

$$\|e^{At}x\| \to 0 \qquad \text{as } t \to \infty .$$

In other words, a state is stable if the zero-input response decays asymptotically to zero. The class of stable states is clearly a linear subspace. Let us denote the class by Σ_S. The orthogonal complement of Σ_S, denoted Σ_U, will be called the "unstable state" (or "A-unstable subspace"). Every element of Σ_U is unstable, but not every unstable state needs be in Σ_U.

Let A be simple. Then Σ_S is spanned by those eigenvectors with eigenvalues whose real parts are negative. The remaining eigenvectors have nonzero projections on Σ_U.

Problems for Chapter 5

1. Construct a matrix A such that A is stable but for *some* $t > 0$

$$\|e^{At}\|_o > 1 .$$

Hint: Use

$$A = \begin{vmatrix} 0 & 1 \\ -\frac{1}{4} & 0 \end{vmatrix} .$$

Can you come up with a general construction?

2. Construct (if possible) two stable matrices whose sum is not stable.

3. Show that the solution $x(t)$ of

$$\dot{x}(t) = Ax(t)$$

is such that

$$\|x(t)\|^2 = \|x(a)\|^2 , \qquad t \geq a ,$$

if and only if

$$A + A^* = 0 .$$

4. Find Σ_U and Σ_S for

$$A = \begin{vmatrix} -2 & 0 & 1 & -2 \\ 0 & 0 & 1 & 0 \\ 1 & 0 & -3 & 1 \\ 0 & 0 & 0 & -1 \end{vmatrix} .$$

Find the projection of each of the eigenvectors corresponding to the eigenvalues with nonnegative real parts on the stable subspace.

5. Let A be $n \times n$ and stable. Let C be $m \times n$. Define the $2n \times 2n$ matrix

$$\tilde{A} = \begin{vmatrix} A & I \\ 0 & 0 \end{vmatrix},$$

where I is the Identity $n \times n$ matrix.

Let

$$\tilde{C} = [\, C \quad 0 \,]$$

so that $\tilde{C}$ is $m \times 2n$. Determine necessary and sufficient conditions for $(\tilde{C} \sim \tilde{A})$ to be Observable.

CHAPTER 6

MORE ON CONTROLLABILITY: CANONICAL FORMS

Just as we decomposed a state space into observable and unobservable spaces, we can also obtain a similar decomposition with respect to Controllability. We can also exploit duality in the process. To be specific, let A be $n \times n$ and let B be $n \times q$.

Thus let Σ_c denote the subspace of states reachable from the origin in some time. Recall (Chapter 2) that we have

$$\Sigma_c = \Omega(t) = \bigcup_t \Omega(t) = \text{range of } R(t) = \text{range of } L_c \ .$$

Let Σ_{uc} denote the orthogonal complement of Σ_c. We shall call Σ_{uc} the "uncontrollable" subspace and Σ_c the "controllable" subspace; or sometimes, the "uncontrollable part" and the "controllable part", respectively, of the state space. Since Σ_c is the range of L_c, Σ_{uc} is the null space of L_c^*. Thus

$$\Sigma_{uc} = [x \mid B^* x = 0, \ldots, B^* A^{*^{n-1}} x = 0] \, .$$

In other words, Σ_{uc} is the set of column vectors perpendicular to every column vector in the matrix L_c. Hence Σ_c, the orthogonal complement of Σ_{uc}, is spanned by the linearly independent columns of Σ_c. If the rank of L_c is r, there are r such linearly independent columns (or rows) in L_c. Now we have

THEOREM 6.1. There is a smallest integer $p \leq n$ such that the matrices B, AB, ..., $A^{p-1}B$ are linearly independent, and B, AB, ..., $A^{p-1}B$, $A^{p+k}B$ are linearly dependent for every $k \geq 0$.

PROOF. We have only to invoke the "dual" result we have already proved for $\{A^{*j}C^*\}$ in (4.22).

COROLLARY. Suppose B is $n\times 1$. Then

$$p = r .$$

The corollary thus distinguishes the key property of single input systems.

Let P_c denote the projection corresponding to Σ_c. Then we can readily verify that

$$e^{At}\,\Sigma_c \subset \Sigma_c \tag{6.1}$$

for every $t \geq 0$. Hence for any x:

$$AP_c x = P_c A P_c x .$$

Furthermore, for any u in R^q

$$Bu \in \Sigma_c$$

in the range of L_c, and, of course, so does also

$$e^{At}Bu .$$

Hence for every t

$$\int_0^t e^{A(t-\sigma)}Bu(\sigma)\, d\sigma \in \Sigma_c .$$

Let

$$x_c \in \Sigma_c .$$

Then so does

$$x(t) = e^{At}x_c + \int_0^t e^{A(t-\sigma)}Bu(\sigma)\, d\sigma , \qquad t \geq 0 , \tag{6.2}$$

hence so does

$$\dot{x}(t) .$$

Hence

$$\dot{x}(t) = P_c\dot{x}(t) = P_c[Ax(t) + Bu(t)] = P_c A P_c x(t) + Bu(t)$$

or $x(t)$ satisfies

$$\left.\begin{aligned} \dot{x}(t) &= P_c A P_c x(t) + Bu(t) , \\ x(0) &= P_c x(0) . \end{aligned}\right\} \tag{6.3}$$

Hence given any initial state at $t = 0$, we can decompose the state response as

$$x(t) = e^{At}(x - P_c x) + x_c(t) , \tag{6.4}$$

where

$$\left.\begin{aligned} \dot{x}_c(t) &= (P_c A P_c) x_c(t) + Bu(t) , \\ x_c(0) &= P_c x , \end{aligned}\right\} \tag{6.5}$$

and

$$P_c x_c(t) = x_c(t) .$$

Unfortunately,

$$P_c x(t) \neq x_c(t) ,$$

since, in general,

$$P_c e^{At} x \neq e^{At} P_c x .$$

Nevertheless (6.4) shows that if at any time the state is controllable, then it is controllable for all time thereafter and satisfies (6.5). It is natural to call (6.5) the state equation for the "controllable part" of the system. Moreover, we can give it a "canonical" description for the single input case.

Single Input System

Let us now specialize to the case where B is $n\times 1$. Then $p = r$, and Σ_c is spanned by the column vectors

$$B, AB, ..., A^{r-1}B$$

and

$$A^r B = \sum_0^{r-1} a_j A^j B .$$

Hence every element of Σ_c can be expressed

$$x_c = \sum_0^{r-1} c_j A^j B = Ez ,$$

where

$$z = \begin{vmatrix} c_0 \\ \vdots \\ c_{r-1} \end{vmatrix} ;$$

and E maps R^r onto Σ_c and E is 1:1. Next let us calculate

$$Ax_c = \sum_0^{r-1} c_j A^{j+1} B = \sum_0^{r-1} c_{j-1} A^j B + c_{r-1} A^r B .$$

But

$$A^r B = \sum_0^{r-1} a_j A^j B ,$$

and hence

$$Ax_c = \sum_0^{r-1} c_{j-1} A^j B + c_{r-1} \sum_0^{r-1} a_j A^j B .$$

Hence

$$E^{-1}(Ax_c) = \mathfrak{A}(E^{-1} x_c) ,$$

where

$$\mathfrak{A} = \begin{vmatrix} 0 & 0 & 0 & \cdots & 0 & a_0 \\ 1 & 0 & 0 & \cdots & 0 & a_1 \\ 0 & 1 & 0 & \cdots & 0 & a_2 \\ \vdots & \vdots & \vdots & \ddots & \vdots & \vdots \\ 0 & 0 & 0 & \cdots & 1 & a_{r-1} \end{vmatrix} . \tag{6.6}$$

Next,

$$B \in \Sigma_c$$

and

$$E^{-1}[B] = \begin{vmatrix} 1 \\ 0 \\ \vdots \\ 0 \end{vmatrix} = \mathfrak{B} . \tag{6.7}$$

Hence

$$x_c(t) = Ez(t) ,$$

where

$$\left.\begin{aligned} \dot{z}(t) &= \mathfrak{A}z(t) + \mathfrak{B}u(t) , \\ z(0) &= E^{-1}x_c . \end{aligned}\right\} \tag{6.8}$$

Also for

$$\mathrm{v} = Cx_c$$

we obtain

$$\mathrm{v} = \sum_0^{r-1} c_j CA^j B = |CB, ..., CA^{r-1}B|z . \tag{6.9}$$

Thus (6.8) is a "canonical" representation for the controllable part of the state space, with output given by (6.9).

If instead of the state description, we begin with the input-output relation (2.9), which we rewrite as

$$\frac{d^n \mathrm{v}(t)}{dt^n} - \sum_0^{n-1} a_k \frac{d^k \mathrm{v}(t)}{dt^k} = \sum_0^n b_k \frac{d^k u(t)}{dt^k} , \tag{6.10}$$

where we use n in place of r in (2.9); $\mathrm{v}(\cdot)$ is $m \times 1$ and $\{b_k\}$ are $m \times 1$. We can deduce the (observable) state-space description (2.14):

$$\begin{aligned} \dot{x} &= Ax + \tilde{B}u , \\ \mathrm{v} &= Cx + Du , \end{aligned} \tag{6.11}$$

where A is the compound companion matrix ($mn \times mn$)

$$A = \begin{vmatrix} 0 & I & 0 & \cdots & 0 \\ 0 & 0 & I & \cdots & 0 \\ \vdots & \vdots & \vdots & & \vdots \\ 0 & 0 & 0 & \cdots & I \\ a_0 I & a_1 I & a_2 I & \cdots & a_{n-1} I \end{vmatrix} , \tag{6.12}$$

$$C = \text{row} \mid I, 0, ..., 0 \mid , \tag{6.13}$$

$$\mathfrak{B} = \sum_0^n A^j B b_j , \qquad B = \begin{vmatrix} 0 \\ 0 \\ \vdots \\ I \end{vmatrix} ,$$

$$D = b_n .$$

The minimal polynomial of A is

$$\psi(\lambda) = \lambda^n - \sum_0^{n-1} a_k \lambda^k .$$

Hence (Cayley-Hamilton theorem):

$$A^n = \sum_0^{n-1} a_k A^k$$

so that

$$\sum_0^{n-1} a_k A^k \widetilde{B} = A^n \widetilde{B} .$$

Since $\widetilde{B}$ is a column vetor $(mn \times 1)$ it follows that

$$\dim \Sigma_c \leq n ,$$

however large m may be!

We can say more in the SI-SO case.

Single Input-Single Output System

Let $m = 1$ in (6.10) and let

$$D(\lambda) = \lambda^n - \sum_0^{n-1} a_k \lambda^k ,$$

$$N(\lambda) = \sum_0^n b_k \lambda^k .$$

Then we can go beyond Theorem 3.6 to assert:

THEOREM 6.2.

$$\dim \Sigma_c = n - q = p ,$$

where q is the degree of the highest degree polynomial (Greatest Common Divisor, GCD) which divides $N(\lambda)$ and $D(\lambda)$.

PROOF. We may write the GCD in the form

$$G(\lambda) = \lambda^q + \sum_0^{q-1} c_k \lambda^k .$$

Since $m = 1$, $D(\lambda)$ is the same as $\psi(\lambda)$. Let

$$Q(\lambda) = \psi(\lambda)/G(\lambda) = \lambda^{n-q} + \sum_0^{n-q-1} d_k \lambda^k .$$

Now

$$Q(A)\widetilde{B} = Q(A)N(A)B .$$

Now

$$Q(\lambda)N(\lambda) = \psi(\lambda) \left[\frac{N(\lambda)}{G(\lambda)}\right]$$

and $\psi(\lambda)$ divides $Q(\lambda)N(\lambda)$, and hence

$$Q(A)\widetilde{B} = 0 ,$$

or

$$A^{n-q}\widetilde{B} = - \sum_0^{n-q-1} d_k A^k \widetilde{B} .$$

Hence

$$p \leq n - q .$$

Next, p being by definition the dimension of Σ_r, we must have

$$A^p \widetilde{B} = \sum_0^{p-1} a_k A^k \widetilde{B} \qquad \text{for some } \{a_k\}$$

or, defining

$$\phi(\lambda) = \lambda^p - \sum_0^{p-1} a_k \lambda^k ,$$

we have that

$$0 = \phi(A)\widetilde{B} = \phi(A)N(A)B .$$

Now because for our choice of B, the vectors B, ..., $A^{n-1}B$ are linearly independent, and we can show that $\psi(\lambda)$ must divide $\phi(\lambda)N(\lambda)$. For, suppose not. Then

$$\phi(\lambda)N(\lambda) = H(\lambda)\psi(\lambda) + r(\lambda)$$

and hence

$$\phi(A)N(A)B = \psi(A)H(A)B + r(A)B = r(A)B = 0 .$$

The degree of $r(A)$ being less than n, this would imply that B, ..., $A^{n-1}B$ are *not* linearly independent. Hence $\psi(\lambda)$ divides $\phi(\lambda)N(\lambda)$. Hence

$$\phi(\lambda)N(\lambda) = H(\lambda)\psi(\lambda) .$$

Hence

$$\frac{N(\lambda)}{H(\lambda)} = \frac{\psi(\lambda)}{\phi(\lambda)} = S(\lambda) .$$

Hence degree of the polynomial $S(\lambda)$ must be less than q. But the degree of

$$\frac{\psi(\lambda)}{\phi(\lambda)}$$

must be $(n - p)$ and hence

$$n - p \leq q$$

or

$$p \geq n - q .$$

Hence

$$p = n - q .$$

Next let us go on to find the canonical representation for the controllable part of the system. Let $G(\lambda)$ be the GCD of $\psi(\lambda)$ and $N(\lambda)$ and let, as before,

$$Q(\lambda) = \frac{\psi(\lambda)}{G(\lambda)} = \lambda^p - \sum_0^{p-1} d_k\lambda^k .$$

Then

$$A^p\tilde{B} - \sum_0^{p-1} d_kA^k\tilde{B} = 0$$

and $A^k\tilde{B}$ are linearly independent for $k \leq p-1$. We may therefore proceed as before letting $p = r$ to obtain $\mathcal{Z}$ in (6.6) with d_k replacing a_k, and go on to obtain (6.8). In (6.9) we note that in our case

$$| C\tilde{B} \;\; CA\tilde{B} \;\; \cdots \;\; CA^{r-1}\tilde{B} |$$

can be calculated from

$$CA^k\tilde{B} = \sum_0^n b_j CA^{j+k}B ,$$

noting that

$$\begin{aligned} CA^kB &= 0 , & k < n-1 , \\ &= 1 , & k = n-1 , \\ &= a_{n-1} , & k = n . \end{aligned}$$

Thus

$$\begin{aligned} C\tilde{B} &= b_n a_{n-1} + b_{n-1} , \\ CA\tilde{B} &= b_n a_{n-2} + b_{n-2} , \end{aligned}$$

etc.

Problems For Chapter 6

1. Show that the system with input-output relation

$$\frac{d^3v}{dt^3} + 6\frac{d^2v}{dt^2} + 11\frac{dv}{dt} + 6v(t) = \frac{du}{dt} + 4u(t)$$

can be described in terms of a Controllable and Observable state space in the form

$$\dot{z}(t) = Az(t) + bu(t) ,$$

$$v(t) = Cz(t) ,$$

where

$$b = \begin{vmatrix} 1 \\ 0 \\ 0 \end{vmatrix}$$

and the Controllability matrix is the Identity.

2. For the system with input-output relation

$$\frac{d^3v}{dt^3} - v(t) = \frac{du}{dt} - u(t)$$

determine an Observable state-space representation. Show that the state space is not Controllable. Find a representation for the Controllable part. Find an unreachable state which is not in the Uncontrollable subspace Σ_{uc}. Characterize the subspace reached one second after applying the zero input, starting with Σ_{uc}.

3. Let

$$A = \begin{vmatrix} 2 & -10 & 20 \\ -10 & -13 & -10 \\ 20 & -10 & 2 \end{vmatrix} ;$$

$$C = \begin{vmatrix} 2 & -1 & 2 \\ 2 & 2 & -1 \end{vmatrix} .$$

Characterize:

Stable subspace Σ_s

Unstable subspace

Unobservable subspace Σ_0

Observable subspace $\Sigma_0 \cap \Sigma_s$.

4. Given the state-space description

$$\dot{x} = Ax + Bu ,$$

$$y = Cx ,$$

where

$$A = \begin{vmatrix} 2 & 1 & 0 \\ -2 & -1 & 2 \\ 1 & 1 & 1 \end{vmatrix} ; \qquad B = \begin{vmatrix} 1 \\ 0 \\ 1 \end{vmatrix} ;$$

$$C = | -1 \quad -1 \quad 1 | .$$

Find $\Sigma_s \cap \Sigma_{uc}$. Find the subspace of Unobservable states which are Stable. Find the subspace of Stable states. Find the Unstable state space.

5. Explain what happens when in the construction for the Controllable part in Theorem 6.2, $q = 0$ (no pole-zero cancellation).

CHAPTER 7

STATE ESTIMATION AND STATE STABILIZABILITY

We shall discuss two related topics in this chapter: one is the problem of estimating the state from the input-output data and the other is the problem of stabilizing our otherwise unstable system by "state feedback."

State Estimation

Let us begin with the state estimation problem. Thus we are given a time-invariant linear dynamic system with the state-space description

$$\begin{aligned} \dot{x}(t) &= Ax(t) + Bu(t) , \\ \mathrm{v}(t) &= Cx(t) . \end{aligned} \tag{7.1}$$

The problem is to estimate the state $x(t)$ from $(u(t), \mathrm{v}(t))$. We assume that $u(t)$ and $\mathrm{v}(t)$ are both made available without error. The state at time $t = 0$ satisfies the equation, for any $t_0 > 0$,

$$\int_0^{t_0} e^{A^*s} C^* C e^{As}\, ds \; x(0) = \int_0^{t_0} e^{A^*s} C^* (\mathrm{v}(s) - Cx_u(s))\, ds , \tag{7.2}$$

where

$$\begin{aligned} \dot{x}_u(s) &= Ax_u(s) + Bu(s) , \\ x_u(0) &= 0 . \end{aligned}$$

This equation has a unique solution, only if the state space is observable. But it has some solution, nevertheless. Neglecting the time taken to calculate a solution, we can, if the input-output data is stored, go back to the input data and recreate $x(t)$ as the solution to

$$\dot{x}(t) = Ax(t) + Bu(t) , \qquad t \geq 0 ,$$

$x(0)$ given by (7.2). But this entails a fixed delay of at least t_0, and requires "off-line" calculation.

An alternate and "on-line" procedure† is to define an estimate $\hat{x}(t)$ of the state at time t by means of the dynamic equations:

$$\dot{\hat{x}}(t) = A\hat{x}(t) + Bu(t) + PC^*[\mathrm{v}(t) - C\hat{x}(t)] ,$$

where the matrix P is suitably determined. We may set

$$\hat{x}(0) = 0$$

(or any reasonable "guess" for the state). Let $e(t)$ denote the error

$$e(t) = x(t) - \hat{x}(t) .$$

Then

$$\begin{aligned} \dot{e}(t) &= A(x(t) - \hat{x}(t)) - PC^*(\mathrm{v}(t) - C\hat{x}(t)) \\ &= (A - PC^*C)(x(t) - \hat{x}(t)) , \end{aligned}$$

substituting

$$\mathrm{v}(t) = Cx(t) .$$

Or, the error $e(t)$ satisfies

$$\dot{e}(t) = (A - PC^*C)e(t) ,$$

hence

$$e(t) = e^{(A-PC^*C)t}(x(0) - \hat{x}(0)) .$$

We observe that if P is so chosen that

$$A - PC^*C$$

† This is the so-called "observer" theory, for an elaboration of which see [2].

is stable, then $e(t)$ will go to zero as t goes to infinity. Hence asymptotically at least the error will be zero. The problem then is to choose P. In fact, such a choice is always possible if $(C \sim A)$ is observable.

Single Input-Single Output Case

Let us consider the SI-SO case, with the input-output relation

$$\frac{d^n v(t)}{dt^n} - \sum_{0}^{n-1} a_j \frac{dv^j(t)}{dt^j} = u(t) .$$

Then, as we saw in Chapter 4, we can represent this in the state-space form

$$\dot{z} = Az + \tilde{b}u ,$$

$$v = Cz ,$$

where

$$C = \text{row} \mid 0 \; 0 \; \cdots \; 1 \mid ;$$

$$A = \begin{vmatrix} 0 & 0 & 0 & 0 & 0 & \cdots & 0 & a_0 \\ 1 & 0 & 0 & 0 & 0 & \cdots & 0 & a_1 \\ 0 & 1 & 0 & 0 & 0 & \cdots & 0 & a_2 \\ 0 & 0 & 1 & 0 & 0 & \cdots & 0 & a_3 \\ \vdots & \vdots & \vdots & \vdots & \vdots & \ddots & \vdots & \vdots \\ 0 & 0 & 0 & 0 & 0 & \cdots & 1 & a_{n-1} \end{vmatrix} .$$

Our state-estimate equation would be

$$\dot{\hat{z}}(t) = A\hat{z}(t) + \tilde{b}u(t) + KC^* [v(t) - C\hat{z}(t)] .$$

Hence we need to prove that we can find K so that

$$A - KC^* C$$

is stable. First note that

$$C^* C = \begin{vmatrix} 0 & 0 & 0 & \cdots & 0 & 0 \\ 0 & 0 & 0 & \cdots & 0 & 0 \\ 0 & 0 & 0 & \cdots & 0 & 0 \\ \vdots & \vdots & \vdots & \ddots & \vdots & \vdots \\ 0 & 0 & 0 & \cdots & 0 & 1 \end{vmatrix} .$$

The matrix K is $n\times n$ and hence let

$$K = \{k_{ij}\}, \qquad i,j=1,\dots,n.$$

Then

$$A - KC^*C = \begin{vmatrix} 0 & 0 & 0 & 0 & \cdots & 0 & (a_0-k_{1n}) \\ 1 & 0 & 0 & 0 & \cdots & 0 & (a_1-k_{2n}) \\ 0 & 1 & 0 & 0 & \cdots & 0 & (a_2-k_{3n}) \\ 0 & 0 & 1 & 0 & \cdots & 0 & (a_3-k_{4n}) \\ \vdots & \vdots & \vdots & \vdots & \ddots & \vdots & \vdots \\ 0 & 0 & 0 & 0 & \cdots & 1 & (a_{n-1}-k_{nn}) \end{vmatrix}.$$

The characteristic polynomial is

$$\lambda^n + (a_{n-1} - k_{nn})\lambda^{n-1} + \cdots + (a_0 - k_{1n}).$$

Hence we can certainly choose $\{k_{in}\}$ so as to make the coefficients to have any values we wish and in particular make all the roots have negative real parts. Indeed, we can choose them so that we can place the roots anywhere in the complex plane that we wish. In other words, we have "arbitrary pole-placement capability."

Let E be any nonsingular matrix (transformation). Then if we have a 1:1 transformation of the state space

$$x(t) = Ez(t),$$

we have

$$\hat{x}(t) = E\hat{z}(t).$$

We may of course work directly with the representation in terms of $\hat{x}(\cdot)$:

$$\dot{\hat{x}}(t) = EAE^{-1}\hat{x}(t) + EBu(t) + EKC^*[v(t) - CE^{-1}\hat{x}(t)]$$

since

$$EAE^{-1} - EKC^*CE^{-1} = E[A - KC^*C]E^{-1},$$

and hence is stable.

General Case

Let us consider next the general case. Assume that $(C \sim A)$ is observable. Let $\sigma > 0$ be large enough so that

$$\text{Re [eigenvalues of } (\sigma I + A)] > 0 .$$

Define

$$R = \int_0^\infty e^{-(\sigma I + A)t} C^* C e^{-(\sigma I + A)t} \, dt . \tag{7.3}$$

Then R is of course a covariance, but it is nonsingular by our assumption of $(C \sim A)$ observability. Moreover, R satisfies (and is the unique solution of)

$$-(\sigma I + A^*)R - R(\sigma I + A) + C^* C = 0 . \tag{7.4}$$

Let

$$P = R^{-1} .$$

Then multiplying on the left and right of (7.4) by P, we see that P satisfies

$$-P(\sigma I + A^*) - (\sigma I + A)P + PC^* CP = 0 ,$$

which we can rewrite as

$$(\sigma I + A - PC^* C)P + P(\sigma I + A - PC^* C)^* + PC^* CP = 0 . \tag{7.5}$$

Let us show now that

$$A - PC^* C$$

is stable. Actually we shall show that

$$\sigma I + A - PC^* C$$

is stable. Let λ be an eigenvalue of the adjoint

$$(\sigma I + A - PC^* C)^* x = \lambda x , \qquad x \neq 0 . \tag{7.6}$$

Then substituting in (7.5) and taking inner products,

$$[(\sigma I + A - PC^* C)Px,\ x] + [P(\sigma I + A - PC^* C)^* x,\ x] + [PC^* CPx,\ x]$$
$$= 2(\text{Re}\ \lambda)[Px,\ x] + \|CPx\|^2 = 0 .$$

Hence

$$\operatorname{Re} \lambda = \frac{-\|CPx\|^2}{2[Px,x]}$$

and hence

$$\leq 0 .$$

Suppose

$$\operatorname{Re} \lambda = 0 .$$

Then

$$CPx = 0 .$$

But substituting in (7.6), we have

$$(\sigma I + A^*)x = \lambda x ,$$

which is impossible, since by assumption $(\sigma I + A)$, and hence $(\sigma I + A^*)$ has all eigenvalues with positive real parts.

REMARK. Note that we can choose σ as large as we wish such that

$$\operatorname{Re}\,[\text{eigenvalues of } (A - PC^* C)] \leq -\sigma .$$

In other words, we can make $(A - PC^* C)$ as stable as we wish and thus make $\hat{x}(\cdot)$ approach $x(\cdot)$ as fast as we want.

Stabilizability

Let us look at the related but more important problem of stabilizing an otherwise unstable system by "state feedback"—which is one of the central problems of control theory. Thus let the state equation be

$$\dot{x}(t) = Ax(t) + Bu(t) .$$

We want to choose a "feedback" control of the form

$$u(t) = Kx(t) ,$$

where we assume that we do have the full state $x(t)$ available (estimated from the output, for example). The resulting state equation is

$$\dot{x}(t) = (A + BK)x(t) ,$$

and the problem is to determine K so that the resulting system, i.e.,

$$(A + BK)\ ,$$

is stable. Equivalently,

$$(A^* + K^* B^*)$$

must be stable. But we have just seen that a sufficient condition for this is that $(B^* \sim A^*)$ is observable or $(A \sim B)$ is controllable.

Single Input-Single Output System

Let us specialize to a SI-SO system and see how to construct the "feedback gain matrix K," assuming Controllability. Thus let the input-output relation be

$$\frac{d^n v(t)}{dt^n} - \sum_0^{n-1} a_k \frac{dv^k(t)}{dv(t)^k} = u(t)\ ,$$

with the state-space representation (6.11)

$$\dot{x}(t) = Ax(t) + Bu(t)\ ,$$

where A is defined by (6.12) and

$$B = \begin{vmatrix} 0 \\ 0 \\ 0 \\ \vdots \\ 1 \end{vmatrix} .$$

Let

$$u(t) = -B^* Jx(t)\ .$$

Then

$$A - BB^*J = A - \begin{vmatrix} 0 & 0 & \cdots & 0 \\ \vdots & \vdots & & \vdots \\ 0 & 0 & \cdots & 1 \end{vmatrix} J$$

$$= A - \begin{vmatrix} 0 & 0 & \cdots & 0 \\ \vdots & \vdots & & \vdots \\ 0 & 0 & \cdots & 0 \\ j_{n1} & j_{n2} & \cdots & j_{nn} \end{vmatrix} ,$$

where

$$J = \{j_{ik}\}, \qquad i, k = 1, \ldots, n.$$

Hence

$$\begin{vmatrix} 0 & 1 & 0 & \cdots & 0 \\ \vdots & \vdots & \vdots & & \vdots \\ 0 & 0 & 0 & \cdots & 1 \\ (a_0 - j_{n1}) & (a_1 - j_{n2}) & (a_2 - j_{n3}) & \cdots & (a_{n-1} - j_{nn}) \end{vmatrix},$$

with characteristic polynomial

$$\Delta(\lambda) = \lambda^n + (a_{n-1} - j_{nn})\lambda^{n-1} + \cdots + (a_0 - j_{n1}).$$

Hence we can choose J to place the zeros of $\Delta(\lambda)$ (or the "poles" of the system) anywhere we wish, and in particular to assure stability. Under any 1:1 linear transformation of the state space, we note that

$$EAE^{-1} - EBB^*JE^{-1} = E[A - BB^*J]E^{-1}$$

is also stable.

General Case

Suppose $(A \sim B)$ is controllable. In view of what we have already proved, let us exploit duality. Thus

$$(B^* \sim A^*)$$

is observable. Hence we can find P such that

$$\text{Re [eigenvalues } (A^* - PBB^*)] \leq -\sigma$$

for $\sigma > 0$ and arbitrary. The same is then true for the adjoint

$$(A - BB^*P),$$

as required. Note that

$$P^{-1} = \int_0^\infty e^{-(\sigma I + A)t} BB^* e^{-(\sigma I + A^*)t} \, dt$$

and P of course satisfies

$$P(\sigma I + A) + (\sigma I + A^*)P - PBB^*P = 0.$$

Robust Stabilizers

The stabilizing controller design we have discussed so far depends on the knowledge of the (system) matrices A and B. In many cases these may be known only approximately, and so it is of interest to design controllers which are insensitive to variations from the design values. We refer to them as "robust" controllers. Although robustness is difficult to achieve in general, there is one important class of systems where it is possible to construct stabilizers which do not require quantitative knowledge of the system parameters.

We begin with a general result of this kind. Let the state-space representation be

$$x(t) = Ax(t) + Bu(t)$$

and suppose

(i) $(A + A^*)$ is negative definite, or $-(A + A^*)$ is a covariance:

$$[(A + A^*)x, x] \leq 0 \qquad \text{for every } x,$$

(ii) $(A \sim B)$ is controllable.

Then we shall show that the state feedback control:

$$u(t) = -B^* x(t)$$

is a stabilizing control. In other words, the feedback matrix P can be taken to be the identity matrix:

$$(A - BB^*) \text{ is stable}.$$

To prove this, suppose λ is an eigenvalue of $(A - BB^*)$:

$$Ax - BB^*x = \lambda x, \quad x \neq 0. \tag{7.7}$$

Let

$$\sigma = \text{Re}\,\lambda .$$

Then from (7.7) we have

$$[Ax, x] - [BB^*x, x] = \lambda[x, x]$$

and taking real parts on both sides, we have

$$\text{Re}\,[Ax, x] - [B^*x, B^*x] = \sigma[x, x] .$$

Since

$$2 \operatorname{Re} [Ax, x] = [Ax, x] + \overline{[Ax, x]} = [Ax, x] + [A^*x, x]$$
$$= [(A+A^*)x, x]$$

and hence

$$\operatorname{Re} [Ax, x] \leq 0 ,$$

it follows from the negative definiteness that

$$\sigma \leq 0 .$$

Suppose

$$\sigma = 0 .$$

Then

$$|\operatorname{Re} [Ax, x]| + [B^*x, B^*x] = 0$$

and hence

$$B^*x = 0 .$$

But then

$$Ax - BB^*x = Ax = \lambda x . \tag{7.8}$$

We have so far not exploited the $(A \sim B)$ controllability property. Let us do so now. First of all since $(A - BB^*)$ has an eigenvalue whose real part is zero, $(A^* - BB^*)$ must have an eigenvalue with the same property. Let therefore

$$(A^* - BB^*)z = \lambda z ; \qquad z \neq 0; \ \operatorname{Re} \lambda = 0 .$$

Then we must have

$$\operatorname{Re} [A^*z, z] = [B^*z, B^*z] ,$$

and since

$$2 \operatorname{Re} [A^*z, z] = [(A^*+A)z, z] \leq 0$$

by assumption (i), it follows that

$$B^*z = 0 ,$$

and as a consequence

$$A^*z = \lambda z ,$$

and hence

$$B^*A^{*k}z = 0 , \qquad \text{for every integer } k \geq 0 .$$

This contradicts the assumed ($A \sim B$) controllability.

An important example of this class occurs in "vibrating" systems in Mechanics (see [Chapter 4 in [4]), where the $n \times 1$ vector $x(t)$ denoting linear and/or angular displacements we have the dynamic equations:

$$M\ddot{x}(t) + D\dot{x}(t) + Ax(t) = Bu(t) . \tag{7.9}$$

Here

M (the inertia matrix) is a Covariance ,

D (the viscous damping matrix) is a Covariance ,

A (the stiffness matrix) is a Covariance .

M will be nonsingular but A and/or D may be singular. To obtain the state-space representation, let

$$z(t) = \begin{vmatrix} x(t) \\ \dot{x}(t) \end{vmatrix} .$$

We have then:

$$\dot{z}(t) = \mathcal{A}z(t) + \mathcal{B}u(t) , \tag{7.10}$$

where

$$\mathcal{A} = \begin{vmatrix} 0 & I \\ -M^{-1}A & -M^{-1}D \end{vmatrix} , \tag{7.11}$$

$$\mathcal{B}u = \begin{vmatrix} 0 \\ M^{-1}Bu \end{vmatrix} . \tag{7.12}$$

The energy in the system is given by

$$E(t) = \frac{1}{2}[Ax(t), x(t)] + \frac{1}{2}[M\dot{x}(t), \dot{x}(t)] . \tag{7.13}$$

What we want to show now is that it is possible (under an appropriate controllability condition) to "stabilize" the system in the sense that the energy $E(t)$ can be made to decrease to zero, whatever the initial values of $x(0)$ and $\dot{x}(0)$, by means of a "robust" state feedback.

Now, the time derivative of the energy

$$\frac{d}{dt}E(t) = [Ax(t), \dot{x}(t)] + [M\ddot{x}(t), x(t)] = [Bu(t), \dot{x}(t)] - [D\dot{x}(t), \dot{x}(t)] ,$$

and hence choosing the control input $u(t)$ to be

$$u(t) = -B^*\dot{x}(t) \tag{7.14}$$

yields

$$\frac{d}{dt}E(t) \leq 0 , \tag{7.15}$$

and hence makes the energy nonincreasing. But since the right side of (7.15) may well be zero, this is not enough to show that the energy $E(t)$ will actually decrease to zero. However this is a robust controller with respect to the system, since only B needs to be known. With this choice of feedback control we have

$$\mathcal{B}u(t) = -Kz(t) , \tag{7.16}$$

where

$$K = \begin{vmatrix} 0 & 0 \\ 0 & M^{-1}BB^* \end{vmatrix} .$$

Suppose that the energy $E(t)$ is zero at some instant of time—make it the "initial" time $t = 0$. Then letting

$$x(0) = z_1 \qquad ; \qquad z = \begin{vmatrix} z_1 \\ z_2 \end{vmatrix} ,$$
$$\dot{x}(0) = z_2$$

we have

$$[Az_1, z_1] + [Mz_2, z_2] = 0 ,$$

and since M is nonsingular, it follows that

$$Az_1 = 0 ,$$

$$z_2 = 0 .$$

Further,

$$(\mathcal{A} - K)z = 0$$

and hence (7.10) has the solution

$$z(t) = z, \qquad t \geq 0,$$

and

$$E(t) = 0, \qquad t \geq 0.$$

We may therefore eliminate such states from consideration. We may thus confine ourselves to the subspace orthogonal to the null space of A, or, in other words, assume A is nonsingular. Let us assume now that A is nonsingular.

It is convenient then to make the following one-to-one transformation on the state space by defining

$$Tz = Y,$$

$$T = \begin{vmatrix} \sqrt{A} & 0 \\ 0 & \sqrt{M} \end{vmatrix}. \tag{7.17}$$

Then T is of course nonsingular, and defining

$$Y(t) = Tz(t),$$

equation (7.10) transforms to

$$\dot{Y}(t) = T\mathcal{A}T^{-1}Y(t) + T\mathcal{B}u(t). \tag{7.18}$$

Note in particular that

$$\frac{1}{2}[Y(t), Y(t)] = E(t).$$

The square of the norm is the same as twice the system energy. Let

$$\tilde{\mathcal{A}} = T\mathcal{A}T^{-1} = \begin{vmatrix} 0 & \sqrt{A}\sqrt{M^{-1}} \\ -\sqrt{M^{-1}}\sqrt{A} & -\sqrt{M^{-1}}\,D\,\sqrt{M^{-1}} \end{vmatrix}.$$

Moreover, using

$$u(t) = -B^*\dot{x}(t) ,$$

we have

$$T\mathfrak{B}u(t) = -TKT^{-1}Y(t)$$

and

$$TKT^{-1} = \mathfrak{B}\mathfrak{B}^* ,$$

with $\mathfrak{B}$ defined by

$$\mathfrak{B}u = \begin{vmatrix} 0 \\ \sqrt{M^{-1}}\, Bu \end{vmatrix} .$$

Hence it is enough to show that

$$\mathfrak{A} - \mathfrak{B}\mathfrak{B}^*$$

is stable. First of all it is readily verified that

$$\frac{1}{2}\,[(\mathfrak{A}+\mathfrak{A}^*)Y,\ Y] = -[\sqrt{M^{-1}}\, D\sqrt{M^{-1}}\, y_2,\ y_2] \leq 0 . \tag{7.19}$$

Rather than verify $(\mathfrak{A} \sim \mathfrak{B})$ controllability, it is easier to imitate the proof of the general result. If $(\mathfrak{A} - \mathfrak{B}\mathfrak{B}^*)$ has an eigenvalue whose real part is zero, so must

$$\mathfrak{A}^* - \mathfrak{B}\mathfrak{B}^* .$$

Also, if

$$\mathfrak{A}^*Y - \mathfrak{B}\mathfrak{B}^*Y = \lambda Y , \qquad \operatorname{Re}\lambda = 0,\ \ \|Y\| \neq 0 ,$$

we must have

$$\mathfrak{B}^*Y = 0 \tag{7.20}$$

and hence

$$\mathfrak{A}^*Y = \lambda Y . \tag{7.21}$$

But (7.20) yields

$$B^*\sqrt{M^{-1}}\, y_2 = 0 \,,$$

where

$$Y = \begin{vmatrix} y_1 \\ y_2 \end{vmatrix} .$$

Since from (7.21)

$$[(\mathfrak{A}+\mathfrak{A}^*)Y,\ Y] = 0 \,,$$

we obtain from (7.19) that

$$\sqrt{M^{-1}}\, D\sqrt{M^{-1}}\, y_2 = 0 \,.$$

Hence

$$\mathfrak{A}^*Y = \begin{vmatrix} -\sqrt{A}\,\sqrt{M^{-1}}\, y_2 \\ \sqrt{M^{-1}}\,\sqrt{A}\, y_1 \end{vmatrix} = \begin{vmatrix} \lambda y_1 \\ \lambda y_2 \end{vmatrix} ,$$

or

$$\sqrt{A}\,\sqrt{M^{-1}}\, y_2 = -\lambda y_1 = -\lambda^2\sqrt{A^{-1}}\,\sqrt{M}\, y_2 \,, \tag{7.22}$$

or

$$\sqrt{M^{-1}}\, A\sqrt{M^{-1}}\, y_2 = -\lambda^2 y_2 \,. \tag{7.23}$$

Hence

$$(\sqrt{M^{-1}}\, B^*\sqrt{M^{-1}})(\sqrt{M^{-1}}\, A^*\sqrt{M^{-1}})^k y_2 = 0 \,,$$

$$\text{for every integer } k \ge 0 \,. \tag{7.24}$$

Here y_2 cannot be zero, since otherwise from (7.22) y_1 will also be zero. In other words, we need that

$$((\sqrt{M^{-1}}\, A\sqrt{M^{-1}}) \sim (\sqrt{M^{-1}}\, B\sqrt{M^{-1}}))$$

must be controllable. We can also state this in another equivalent way.

Suppose

$$Ax = \gamma M x , \qquad \text{for some } \gamma, \ x \neq 0 ,$$

then

$$B^* x$$

cannot be zero. This follows since we can rewrite (7.23) as

$$Ah = (-\lambda^2) M h ,$$

where

$$h = \sqrt{M^{-1}}\, y_2 .$$

Problems for Chapter 7

1. For the SI-SO system with input-output relation

$$\frac{d^2 v}{d t^2} + 6 \frac{dv}{dt} + 5 v(t) = u(t) ,$$

find a feedback-state estimator such that the estimator error $z(t)$ has the exponential bound

$$\| z(t) \| \leq e^{-kt} \| z(0) \| .$$

Use both methods of this chapter.

2. Let the state-space representation be

$$\dot{x}(t) = Ax(t) + Bu(t) ,$$

where

$$A = \begin{vmatrix} 0 & 1 \\ -25 & 0 \end{vmatrix} , \qquad B = \begin{vmatrix} 0 \\ 1 \end{vmatrix} .$$

Find P so that

$$\text{Re [eigenvalues } (A - BB^* P)] \leq -\sigma$$

for $\sigma > 0$ given, by solving

$$(\sigma I + A)R + R(\sigma I + A^*) - BB^* = 0$$

and taking

$$P = R^{-1} .$$

Compare this with solving directly for P from

$$(\sigma I + A^*)P + P(\sigma I + A) - PBB^*P = 0 .$$

3. For the system

$$\dot{x}(t) = Ax(t) + Bu(t) ,$$

with A, B as in Problem 2, and

$$v(t) = Cx(t) ,$$

where C = row [1, 0], construct a cascade-state-estimator-feedback-controller in the form

$$u(t) = -B^* P_c \hat{x}(t) ,$$

$$\dot{\hat{x}}(t) = A\hat{x}(t) + P_f C^* [v(t) - C\hat{x}(t)] + Bu(t) ,$$

where P_f is such that

$$\|\hat{x}(t) - x(t)\| \le \|\hat{x}(0) - x(0)\| e^{-kt}$$

and P_c is such that

$$\text{Re [eigenvalues } (A - BB^* P)] \le -\sigma .$$

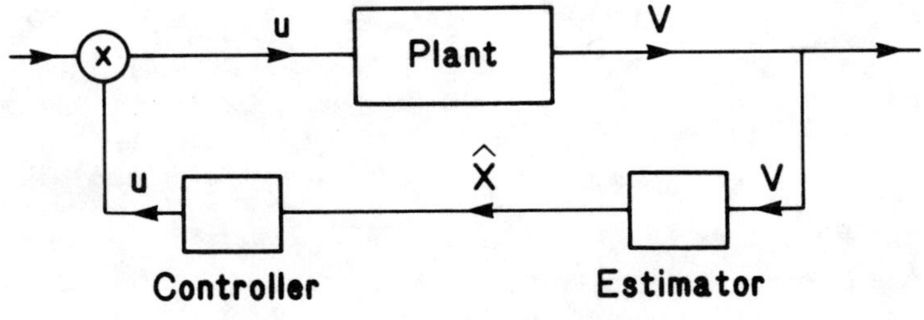

Determine whether the total system is Stable.

Determine whether the feedback block ("compensator") with v(·) as input and $u(\cdot)$ as output is Observable; Controllable; Stable.

References for Chapter 7

[1] Anderson, B.D.O., and Moore, J.B. *Linear Optimal Control.* Englewood Cliffs, N.J.: Prentice-Hall Press, 1971.

[2] Wonham, W.M. *Linear Multivariable Control. A Geometric Approach.* New York Berlin Heidelberg: Springer-Verlag, 1978.

[3] Padullo, L., and Arbib, M.A. *System Theory.* Philadelphia, PA: W.B. Saunders Co., 1974.

[4] Timoshenko, S.P., Young, D.H., and Weaver, W., Jr. *Vibration Problems in Engineering.* 4th ed. New York: John Wiley & Sons Publ. Co, 1974.

CHAPTER 8

TIME-VARYING SYSTEMS

State-Space Description

So far we have considered linear systems which were "time-invariant": the system matrices *A*, *B*, *C*, *D* in the state-space representation were not dependent on time. In many problems we do need to consider linear systems without this restriction. Although there are no new "phenomena" that appear in this generalization, the attendant mathematical complexity is greater, and we shall have to be satisfied with a relatively brief discussion here and leave more detailed study to the references cited.

The simplest time-varying input-output relation for SI-SO systems would be in terms of time-varying linear differential equations. For example,

$$\frac{d^2 \mathrm{v}}{d t^2} + a_1(t)\frac{d\mathrm{v}}{dt} + a_0(t)\mathrm{v}(t) = u(t) . \tag{8.1}$$

Imitating our approach to obtain a state-space representation, we let

$$x(t) = \begin{vmatrix} \mathrm{v}(t) \\ \dfrac{d\mathrm{v}(t)}{dt} \end{vmatrix}$$

and obtain

$$\left.\begin{aligned} \dot{x}(t) &= A(t)x(t) + Bu(t) , \\ \mathrm{v}(t) &= Cx(t) , \end{aligned}\right\} \tag{8.2}$$

where

$$A(t) = \begin{vmatrix} 0 & 1 \\ -a_0(t) & -a_1(t) \end{vmatrix} ;$$

$$B = \begin{vmatrix} 0 \\ 1 \end{vmatrix} ; \qquad C = | 1, 0 | .$$

Hence a general form of state-space description of time-varying linear dynamic system would be

$$\begin{aligned} \dot{x}(t) &= A(t)x(t) + B(t)u(t) , \\ \mathrm{v}(t) &= C(t)x(t) , \end{aligned} \tag{8.3}$$

where, to be specific, $A(t)$ is $n \times n$, $B(t)$ is $n \times p$, $u(t)$ is $p \times 1$, $C(t)$ is $m \times n$. It is not possible, in general, to obtain an equivalent input-output relation in the form of a differential equation like (8.2) even for a SI-SO system. Hence we shall only consider the description (8.3) in what follows.

The first problem is the solution of (8.3). For this purpose we need to digress a bit and recall relevant results from the theory of differential equations.

We begin by considering the case of zero input; in this case we have a "homogeneous" equation

$$\left. \begin{aligned} \dot{x}(t) &= A(t)x(t) , \qquad a < t , \\ x(a) & \text{ given} . \end{aligned} \right\} \tag{8.4}$$

If the dimension is 1 ($n = 1$), then the solution is

$$x(t) = e^{\int_a^t A(s)ds} \, x(a) ;$$

but of course this does not work in higher dimensions. Hence we proceed as in the time-invariant case and consider the $(n \times n)$ matrix equation

$$\left. \begin{aligned} \dot{\Phi}(t) &= A(t)\Phi(t) , \qquad t > a , \\ \Phi(a) & \text{ given} . \end{aligned} \right\} \tag{8.5}$$

From the theory of linear differential equations (see [1], for example) we know that this equation has a unique solution. In fact, we may construct the solution by the following iteration technique (named after Picard):

$$\begin{aligned} \Phi_1(t) &= \Phi(a) , \\ \Phi_{n+1}(t) &= \Phi(a) + \int_a^t A(s)\Phi_n(s) \, ds . \end{aligned}$$

Under the condition that

$$\int_a^{a+L} \|A(s)\|^2 \, ds < \infty ,$$

we can show that the sequence converges for each t, $a \le t \le a+L$, and yields, in the limit, the solution $\Phi(t)$ of (8.5) with the property that

$$\lim_{t\to a} \|\Phi(t) - \Phi(a)\| = 0 .$$

Next let

$$m(t) = \text{Det.}\ (\Phi(t)) .$$

Then:

LEMMA 8.1.

$$\frac{dm(t)}{dt} = (\text{Tr.}\ A(t))\, m(t) , \tag{8.6}$$

$$A(t) = \{a_{ij}(t)\} .$$

PROOF. Let

$$\Phi(t) = \{\phi_{ij}(t)\} .$$

Then

$$\dot{\phi}_{ij}(t) = \sum_{k=1}^{n} a_{ik}(t)\phi_{kj}(t) . \tag{8.7}$$

On the other hand, by definition of the determinant as sums of nonrepeating products of elements of the matrix, we can see that

$$\frac{d}{dt}\text{Det.}\ (\Phi(t)) = \sum_{i=1}^{n} \text{Det.} \begin{vmatrix} \phi_1(t) \\ \vdots \\ \dot{\phi}_i(t) \\ \vdots \\ \phi_n(t) \end{vmatrix}$$

where

$$\phi_i(t) = \text{row}\ |\ \phi_{i1}(t) \quad \phi_{i2}(t) \quad \cdots \quad \phi_{in}(t)\ | .$$

Elementary row operations using (8.7) yield that the sum on the right

$$= \sum a_{ii}(t) \text{ Det. } \Phi(t)$$

$$= (\text{Tr. } A(t))m(t) .$$

COROLLARY.

$$\text{Det. } \Phi(t) = \left(e^{\int_a^t \text{Tr. } A(s) ds} \right) \text{Det. } \Phi(a) .$$

PROOF. The proof is immediate by solving (8.6).

By a "fundamental matrix solution" of (8.5) we mean any solution such that $\Phi(t)$ is nonsingular.

We can now proceed to solve

$$\left.\begin{aligned} \dot{x}(t) &= A(t)x(t) + B(t)u(t) , \\ x(a) &\text{ given} . \end{aligned}\right\} \tag{8.8}$$

We imitate the procedure for the one-dimensional case. First multiply both sides by the inverse of fundamental matrix solution $\Phi(t)$ of (8.5). Then

$$\Phi(t)^{-1} \dot{x}(t) - \Phi(t)^{-1} A(t)x(t) = \Phi(t)^{-1} B(t)u(t) . \tag{8.9}$$

Note now that

$$\frac{d}{dt}(\Phi(t) \, \Phi(t)^{-1}) = \frac{d}{dt}(I) = 0 .$$

On the other hand, by the product-differentiation rule, we have that the left side is

$$= A(t)\Phi(t)\Phi(t)^{-1} + \Phi(t) \frac{d}{dt}\Phi(t)^{-1} .$$

Hence

$$\frac{d}{dt}(\Phi(t)^{-1}) = -\Phi(t)^{-1} A(t) ;$$

hence (8.9) can be written

$$\frac{d}{dt}(\Phi(t)^{-1}x(t)) = \Phi(t)^{-1}B(t)u(t) .$$

Hence, integrating both sides and then multiplying by $\Phi(t)$, we have

$$x(t) = \Phi(t)\Phi(a)^{-1}x(a) + \Phi(t)\int_a^t \Phi(s)^{-1}B(s)u(s)\,ds$$

$$= \psi(t, a)x(a) + \int_a^t \psi(t, s)B(s)u(s)\,ds ,$$

where

$$\psi(t, s) = \Phi(t)\Phi(s)^{-1} . \tag{8.10}$$

We define $\psi(t, s)$ by (8.10) for *all* t and s in $a \le s, t \le a+L$. Note that

$$\psi(t, \tau)\ \psi(\tau, s) = \psi(t, s) , \tag{8.11}$$

which we call the "transition" property, and $\psi(t, s)$ the "transition matrix." Of course,

$$\psi(t, t) = I$$

(independent of the initial value $\Phi(a)$).

Adjoint Equations

Note that the adjoint $\Phi(t)^*$ satisfies

$$\frac{d}{dt}\Phi(t)^* = \Phi(t)^* A(t)^* .$$

Hence the "backward adjoint" equation (occurring in optimal control theory)

$$\left.\begin{aligned} \dot{x}(t) + A(t)^* x(t) &= z(t) , \quad a < t < T , \\ x(T) &= 0 , \end{aligned}\right\} \tag{8.12}$$

can be solved by a similar procedure as we used in connection with (8.8). Multiply both sides by $\Phi(t)^*$. Then (8.12) can be written as

$$\frac{d}{dt}(\Phi(t)^* x(t)) = \Phi(t)^* z(t) .$$

Hence, integrating from t to T, we have

$$\int_t^T \frac{d}{ds}(\Phi(s)^* x(s))\, ds = \int_t^T \Phi(s)^* z(s)\, ds .$$

Since $x(T) = 0$, this yields

$$-\Phi(t)^* x(t) = \int_t^T \Phi(s)^* z(s)\, ds$$

or

$$x(t) = -\int_t^T (\Phi(t)^{*-1}\Phi(s)^*)\, z(s)\, ds ,$$

$$x(t) = -\int_t^T \psi(s, t)^* z(s)\, ds ,$$

where $\psi(t, s)$ is the transition matrix defined by (8.11).

Controllability and Observability

Although more complex than in the time-invariant case (and not quite as useful), the concept of controllability and observability may be extended to time-varying systems. Let us consider controllability.

DEFINITION. The state space of (8.3) is said to be "completely controllable" if given any state x at time a we can find a control $u(\cdot)$ so as to reach any state y in some time interval $[a, a+T]$.

Thus complete controllability means that given any two states x, y, we can find $u(\cdot)$ such that

$$y = \Phi(a+T)\Phi(a)^{-1}x + \int_a^{a+T} \Phi(a+T)\Phi(s)^{-1}B(s)u(s)\, ds$$

for some $T > 0$. In particular, taking $x = 0$, all states must be reachable at some time $a + T$, starting from the zero state at time $t = a$. There are various other notions of controllability for which the references listed should be consulted. Here we shall merely state a sufficient condition for complete controllability.

THEOREM. Assume that $A(t)$, $B(t)$ are $(n-1)$ times continuously differentiable. Then a sufficient condition for complete controllability is that the rank of the $(n\times np)$ "controllability" matrix

$$| B_0(s) \ \cdots \ B_i(s) \ B_{n-1}(s) | ,$$

where

$$B_0(s) = B(s) ,$$

$$B_k(s) = -A(s)B_{k-1}(s) + \frac{d}{ds} B_{k-1}(s) , \qquad k = 1, \ldots, n-1 ,$$

is n (or controllability matrix has full rank) for every s.

PROOF. The sufficient condition ensures that the matrix

$$R(a, a+T) = \int_a^{a+T} \Phi(s)^{-1} B(s)B(s)^* \Phi(s)^{*-1} \, ds$$

is nonsingular for every a and $T > 0$. In fact, suppose for some a and $T > 0$ this is not the case. Then

$$[R(a, a+T)x, \ x] = 0 , \qquad x \neq 0 ,$$

or

$$B(s)^* \Phi(s)^{*-1} x = 0 , \qquad a \leq s \leq a+T .$$

Hence for every u in R^p

$$[x, \ \Phi(s)^{-1} B(s)u] = 0 , \qquad a \leq s \leq a+T .$$

Let us differentiate with respect to s. Recall that

$$\frac{d}{ds} \Phi(s)^{-1} = -\Phi(s)^{-1} A(s) .$$

Hence

$$\left[x, \ \Phi(s)^{-1} (-A(s)B(s) + \frac{d}{dt} B(s))u\right] = 0$$

or

$$[x, \ \Phi(s)^{-1} B_1(s)u] = 0 .$$

By repeated differentiation we obtain that

$$[x,\ \Phi(s)^{-1}B_k(s)u] = 0, \qquad k = 0, \ldots, n-1\,;$$

hence

$$[\Phi(s)^{-1*}x,\ B_k(s)u] = 0$$

since

$$\Phi(s)^{-1*}x \quad \text{is nonzero}\,.$$

This means that the vectors

$$\{B_k(s)u\}\,, \qquad k = 0, \ldots, n-1\,,$$

are linearly dependent for *all* u in R^p. Hence the rank of the controllability matrix is less than n for all s, $a \le s \le a+T$, which is a contradiction. Hence

$$R(a,\, a+T) \quad \text{is nonsingular}$$

for every a and T. Given any x and y, we note that

$$y = \Phi(a+T)\Phi(a)^{-1}x + \int_a^{a+T} \Phi(a+T)\Phi(s)^{-1}B(s)u(s)\,ds$$

is satisfied for fixed a and T, by taking

$$u(s) = B(s)^*\Phi(s)^{-1*}z\,, \qquad a \le s \le a+T\,,$$

where

$$\Phi(a+T)R(a,\, a+T)z = y - \Phi(a+T)\Phi(a)^{-1}x$$

or

$$R(a,\, a+T)z = \Phi(a+T)^{-1}(y - \Phi(a+T)\Phi(a)^{-1}x)\,,$$

which is possible since $R(a,\, a+T)$ is nonsingular.

REMARK. Note that the sufficient condition implies that we can go from any state to any state in *any* time interval T, however small. Also, the condition reproduces our earlier one for the time-invariant case.

Similarly, we say that the state space is "Completely *Observable*" if the zero-input response

$$v(t) = 0, \qquad t \geq a,$$

implies that the state at time $t = a$ is zero. We can again find a sufficient condition for complete observability.

THEOREM. The state space of (8.3) is completely observable if assuming $C(t)$, $A(t)$ are $(n-1)$ times continuously differentiable, the rank of the $mn \times n$ "observability matrix"

$$\left| \begin{matrix} C_0(s) \\ C_1(s) \\ \vdots \\ C_{n-1}(s) \end{matrix} \right| ,$$

where

$$C_k(s) = \frac{d}{ds} C_{k-1}(s) + C_{k-1}(s)A(s),$$

$$C_0(s) = C(s),$$

is n (or it has full rank) for every s.

PROOF. Suppose the zero-input response

$$v(s) = 0, \qquad s \geq a,$$

for some nonzero state x at time a. Then

$$C(s)\Phi(s)x = 0, \qquad s \geq a, \; x \neq 0.$$

By repeated differentiation we readily obtain that

$$C_k(s)\Phi(s)x = 0, \qquad s \geq a, \; k = 0, \ldots, n-1.$$

Since

$$\Phi(s)x \neq 0,$$

this means that the rank of the observability matrix is less than n in $a \leq s$, which is a contradiction.

REMARK. Note that the sufficient condition ensures that the matrix

$$\int_a^{a+T} \Phi(s)^* C(s)^* C(s) \Phi(s) \, ds$$

is nonsingular for every a and T.

Representation of Linear Dynamic Systems

Finally, let us note a simpler looking state-space representation than (8.3) for every linear dynamic system. Thus let

$$Y(t) = \Phi(t)^{-1} x(t) .$$

Then differentiating with respect to t we have

$$\frac{d}{dt} Y(t) = -\Phi(t)^{-1} A(t) x + \Phi(t)^{-1} A(t) x(t) + \Phi(t)^{-1} B(t) u(t)$$

or

$$\frac{d}{dt} Y(t) = \Phi(t)^{-1} B(t) u(t) .$$

Also

$$\begin{aligned} v(t) &= C(t) x(t) \\ &= C(t) \Phi(t) Y(t) . \end{aligned}$$

Hence we obtain the form

$$\left. \begin{aligned} \dot{x}(t) &= B(t) u(t) , \\ v(t) &= C(t) x(t) \end{aligned} \right\} \tag{8.13}$$

for every linear dynamic system. Note that the coefficients in (8.13) are time-varying even if the system is time-invariant!

Problems for Chapter 8

1. Given the representation

$$\dot{x}(t) = B(t)u(t) ,$$

$$v(t) = C(t)x(t) ,$$

and given further that the system is time-invariant. Show how to find a representation in the form

$$\dot{z}(t) = Az(t) + Bu(t) ,$$

$$v(t) = Cz(t) .$$

2. What does the sufficient condition for complete Controllability/Observability become in terms of the representation (8.13)?

3. Given that

$$\dot{x}(t) = Ax(t) + Bu(t) ,$$

$$u(t) = -B^* z(t) ,$$

$$\dot{z}(t) + A^* z(t) = x(t) , \qquad 0 < t < T ,$$

$$z(T) = 0 .$$

Show that $x(t)$ satisfies

$$\dot{x}(t) = Ax(t) - BB^* \int_t^T e^{A^*(s-t)} x(s)\, ds .$$

Show also that $u(t)$ can be expressed

$$u(t) = -B^* P(t) x(t) ,$$

yielding the time-varying equation

$$\dot{x}(t) = (A - BB^* P(t))x(t) .$$

Determine the differential equation $P(t)$ must satisfy.

Hint: The differential equation is

$$\dot{P}(t) + P(t)A + A^* P(t) + I - P(t)BB^* P(t) = 0 .$$

4. This problem illustrates how linear systems may be considered as arising as approximations to nonlinear systems. Thus, let the nonlinear dynamic system be defined by

$$\dot{x}(t) = f(t, x(t), u(t)) ,$$

$$\mathrm{v}(t) = C(t, x(t)) .$$

Let $x_0(t)$ be a "nominal" solution (or "trajectory") corresponding to the input $u_0(t)$, and initial state $x_0(0)$. Let

$$\Delta u(t) = u(t) - u_0(t) ,$$

$$\Delta x(t) = x(t) - x_0(t) ,$$

$$\Delta x(0) = x(0) - x_0(0) .$$

Derive the "approximate" linear equations valid for "small" perturbations, $\Delta u(t)$, $\Delta x(0)$.

SOLUTION:

$$\begin{aligned} \dot{x}(t) &= \dot{x}_0(t) + \Delta\dot{x}(t) \\ &= f(t, x_0(t), u_0(t)) + f(t, x(t), u(t)) - f(t, x_0(t), u_0(t)) . \end{aligned}$$

We expand $f(t, x, u)$ in a Taylor series in x and u, yielding

$$\begin{aligned} f(t, x, u) = f(t, x_0, u_0) &+ \nabla_x f(t, x_0, u_0)(x - x_0) \\ &+ \nabla_u f(t, x_0, u_0)(u - u_0) \\ &+ \text{higher order terms in } (x - x_0) \text{ and } (u - u_0) , \end{aligned}$$

where ∇_x, ∇_u denote the gradients with respect to x and u, respectively. Omitting terms of order higher than linear in Δu, Δx, we get

$$\begin{aligned} &f(t, x(t), u(t)) - f(t, x_0(t), u_0(t)) \\ &\quad = \nabla_x f(t, x_0(t), u_0(t))\Delta x(t) + \nabla_u f(t, x_0(t), u_0(t))\Delta u(t) . \end{aligned}$$

Similarly,

$$\Delta \mathrm{v}(t) = C(t, x(t)) - C(t_0, x_0(t)) = \nabla_x C(t, x_0(t))\Delta x .$$

Hence we have the linear dynamic system representation:

$$\Delta\dot{x}(t) = A(t)\Delta x(t) + B(t)\Delta u(t) ,$$

$$\Delta \mathrm{v}(t) = C(t)\Delta x(t) ,$$

where

$$A(t) = \nabla_x f(t, x_0(t), u_0(t)) ,$$

$$B(t) = \nabla_u f(t, x_0(t), u_0(t)) ,$$

$$C(t) = \nabla_x C(t, x_0(t)) .$$

Note that $A(t)$, $B(t)$, $C(t)$ will be time-varying even if $f(t, x, u)$ and $C(t, x)$ are independent of t—i.e., even if the original dynamic system is time-invariant. Additional assumptions are thus required if the pertubation system is to be time-invariant.

References for Chapter 8

[1] Zadeh, L.A., and Polak, E. *System Theory*. New York: McGraw-Hill Inc., 1969.

[2] Padullo, L., and Arbib, M.A. *System Theory*. Philadelphia, N.J.: W.B. Saunders & Co., 1974.

CHAPTER 9

DISCRETE-TIME SYSTEMS

So far we have considered continuous-time models of linear systems. Often, if digital computation is involved, we need to work with discrete-time versions which arise if the inputs and outputs are "sampled" at some fixed rate (say $1/\Delta$ per second). Thus we have

$$\begin{aligned} \mathrm{v}_n &= \mathrm{v}(n\Delta) , \\ u_n &= u(n\Delta) ; \end{aligned}$$

and we can deduce a "discrete-time" version of (3.1) in the following way. Let

$$x_n = x(n\Delta) .$$

Then solving (3.1), we have

$$\begin{aligned} x_{n+1} &= e^{A\Delta} x_n + \int_{n\Delta}^{(n+1)\Delta} e^{A((n+1)\Delta-\sigma)} Bu(\sigma)\, d\sigma \\ &= e^{A\Delta} x_n + \int_0^{\Delta} e^{A(\Delta-\sigma)} Bu(n\Delta+\sigma)\, d\sigma . \end{aligned}$$

For Δ sufficiently small, the second term may be approximated as

$$\int_0^{\Delta} e^{A(\Delta-\sigma)} Bu(n\Delta+\sigma)\, d\sigma \approx B\Delta u_n .$$

Other approximations are also possible—such as

$$\int_0^\Delta e^{A\sigma}\, d\sigma \times Bu_n\,, \qquad B\Delta \frac{u_{n+1} + u_n}{2}\,.$$

The accuracy of the approximation depends on the class of inputs as well, as the Sampling Principle (cf. [1]) shows. In any event, we have the representation in the form:

$$\left.\begin{aligned} x_{n+1} &= \mathcal{A}x_n + \mathcal{B}u_n\,, \\ v_n &= Cx_n + Du_n\,, \end{aligned}\right\} \tag{9.1}$$

totally analogous to the continuous version.

We may consider (9.1) as the canonical "discrete-time" linear dynamic system in state-space form, whether or not obtained by "sampling" a continuous-time output. We shall also replace $\mathcal{A}$ by A and $\mathcal{B}$ by B in what follows.

Stability, Controllability, and Observability

The concepts of Stability, Controllability, and Observability are easily extended to (9.1); in fact, the considerations are much simpler. We shall now examine these briefly.

First of all we may "solve" (9.1) as

$$\left.\begin{aligned} x_n &= A^n x_0 + \sum_0^{n-1} A^j B u_{n-1-j}\,, \\ v_n &= CA^n x_0 + \sum_0^{n-1} CA^j B u_{n-1-j} + Du_n\,. \end{aligned}\right\} \tag{9.2}$$

The discrete-time system is

STABLE: If $\|A^n x_0\| \to 0$ as $n \to \infty$ for every x_0 in R^n.

CONTROLLABLE: If any state x is reachable from the zero state in a finite number of time steps. Or, more precisely,

$$x = \sum_0^{n-1} A^j B u_{n-1-j}$$

for some n and input sequence $u_0, \ldots, u_{n-1}$.

OBSERVABLE: If

$$CA^j x = 0 \qquad \text{for every } j \geq 0$$

$$\Rightarrow \quad x = 0\,.$$

Let us deduce necessary and sufficient conditions for each of these properties to hold.

THEOREM 9.1. The system with state-space description (9.1) is observable if and only if the matrix

$$R_0 = \sum_0^{n-1} A^{*k} C^* C A^k$$

is nonsingular. (A is $n \times n$.) Or, equivalently, the observability matrix

$$\begin{vmatrix} C \\ CA \\ \vdots \\ CA^{n-1} \end{vmatrix}$$

has full rank. The observable states are the range of R_0.

THEOREM 9.2. The system with state-space description (9.1) is controllable if and only if the controllability matrix

$$| B \ AB \ \cdots \ A^{n-1}B |$$

has full rank. Or, equivalently,

$$R_c = \sum_0^{n-1} A^j BB^* A^{*j}$$

is nonsingular. The controllable states are the range of R_c.

We omit the proofs since they are actually simpler versions of the proofs for the continuous case.

THEOREM 9.3. The system with state-space description (9.1) is stable if and only if the eigenvalues of A are strictly less than one in magnitude (and strictly inside the unit circle in the complex plane).

PROOF. Let e_i be an eigenvector corresponding to the eigenvalue λ_i of A

$$Ae_i = \lambda_i e_i$$

since

$$\|A^n e_i\| = |\lambda_i|^n .$$

The condition is clearly necessary. Suppose

$$|\lambda_i| < 1 \qquad \text{for all } i .$$

If A is simple, then

$$x = \sum_1^n a_i e_i ,$$

where $\{e_i\}$ are the eigenvalues of A, and

$$A^m x = \sum_1^n a_i \lambda_i^m e_i .$$

Hence

$$\|A^m x\| \leq \sum_1^n |a_i| \, |\lambda_i|^m \to 0 \qquad \text{as } m \to \infty .$$

For the more general case, we may use the result proved in Chapter 1, relating the spectral radius to the operator norm

$$r = \max_i |\lambda_i| = \lim_{n \to \infty} (\|A^n\|_o)^{1/n} ,$$

from which it follows that if $r < 1$, we can find γ such that $r < \gamma < 1$ and integer N such that

$$\|A^n\|_o < \gamma^n \qquad \text{for all } n > N .$$

REMARK.

$$\|A^n x\|^2 = \|AA^{n-1}x\|^2 = [(A^*A)A^{n-1}x, \; A^{n-1}x]$$

so that

$$\mu \leq \frac{\|A^n x\|^2}{\|A^{n-1} x\|^2} \leq \gamma ,$$

where

$$\gamma = \text{largest eigenvalue of } A^*A ,$$

$$\mu = \text{smallest eigenvalue of } A^*A .$$

Hence a sufficient condition for stability is that

$$\gamma = \|A^*A\|_o < 1 ,$$

or, equivalently,

$$\|A\|_o < 1 .$$

The example

$$A = \begin{vmatrix} 0 & 1 \\ a & 0 \end{vmatrix}, \qquad 0 \le a < 1 ,$$

where the spectral radius is $\sqrt{a}$ and the operator norm equals 1, shows that the condition is *not* necessary. In this example

$$A\begin{vmatrix} 0 \\ 1 \end{vmatrix} = \begin{vmatrix} 1 \\ 0 \end{vmatrix}, \qquad \text{so that } \|A\|_o = 1$$

while the eigenvalues are $\pm\sqrt{a}$ so that the spectral radius is $\sqrt{a}$. Note that A is simple.

If (9.1) arises by discretization of (3.1) at the sampling rate of $1/\Delta$ per second, we may wish to relate the properties of (9.1) to those of (3.1). First of all in that case A in (9.1) is always nonsingular. Next (9.1) is stable if and only if (3.1) is stable. However, (9.1) may *not* be observable even if (3.1) is. Thus let

$$C = [1, 0] ,$$

$$A = \begin{vmatrix} 0 & 1 \\ -\omega_0^2 & 0 \end{vmatrix} .$$

Then

$$e^{At}\begin{vmatrix} x_1 \\ x_2 \end{vmatrix} = \begin{vmatrix} x_1 \cos \omega_0 t + x_2 \dfrac{\sin \omega_0 t}{t} \\ -x_1 \omega_0 \sin \omega_0 t + x_2 \cos \omega_0 t \end{vmatrix} .$$

If

$$\Delta = \frac{\pi}{\omega_0} ,$$

we see that

$$Ce^{kA\Delta}\begin{vmatrix} 0 \\ 1 \end{vmatrix} = 0 \qquad \text{for every integer } k ,$$

while, of course,

$$Ce^{At}\begin{vmatrix}0\\1\end{vmatrix}$$

is *not* zero for all $t > 0$.

We can readily generalize this result to matrices which are simple. Thus we can state:

THEOREM 9.4. Let A be simple, and $(C \sim A)$ observable. If for some $\Delta > 0$

$$(C \sim e^{A\Delta})$$

is *not* observable, we must necessarily have

$$e^{\lambda_1 \Delta} = e^{\lambda_2 \Delta} \tag{9.3}$$

for two distinct eigenvalues λ_1, λ_2 of A.

PROOF. Not all eigenvalues of A can be the same, since A, being simple, would be a multiple of the identity and the nonobservability of $(C \sim e^{A\Delta})$ would imply that of $(C \sim A)$, contradicting the hypothesis. Let $\{\lambda_i\}$, $i = 1,...,p$, denote the distinct eigenvalues, m_i denoting the dimension of the corresponding eigenfunction space. For each i, let e_{ij}, $j = 1,...,m_i$, denote the eigenvectors:

$$\lambda_i e_{ij} = Ae_{ij}\ , \qquad j = 1,...,m_i\ .$$

Suppose for some nonzero x,

$$Ce^{Ak\Delta}x = 0\ , \qquad \text{for every integer } \geq 0\ .$$

Now x can be expressed as

$$x = \sum_{i=1}^{p} \sum_{j=1}^{m_i} a_{ij} e_{ij}$$

and hence

$$Ce^{Ak\Delta}x = \sum_{i=1}^{p} \left[\sum_{j=1}^{m_i} a_{ij}\, Ce_{ij}\right] e^{\lambda_i k} = \sum_{i=1}^{p} z_i^k r_i\ ,$$

where

$$z_i = e^{\lambda_i \Delta}\ ,$$

$$r_i = \sum_{1}^{m_i} a_{ij} C e_{ij}\ .$$

Suppose now that for every $k\ell$

$$[r_i, Ce_{k\ell}] = 0$$

for some i. Then

$$\left[C^* C \sum_1^{m_i} a_{ij} c e_{ij},\ e_{k\ell}\right] = 0$$

and hence

$$C^* C \sum_1^{m_i} a_{ij} c e_{ij} = 0 ,$$

or

$$r_i = 0 .$$

Since $(C \sim A)$ is observable, we can find $k\ell$ such that not *all* b_i, where

$$b_i = [r_i, Ce_{k\ell}]$$

are zero. Hence

$$\sum_1^p z_i^k b_i = 0 \qquad \text{for every integer } k \geq 0 .$$

Let

$$b = \begin{vmatrix} \bar{b}_1 \\ \vdots \\ \bar{b}_p \end{vmatrix}, \qquad \text{bar denoting conjugate ,}$$

and

$$h_k = \begin{vmatrix} z_1^k \\ \vdots \\ z_p^k \end{vmatrix} .$$

Then we have

$$[h_k, b] = 0 , \qquad k = 0, 1, \ldots, p-1$$

and hence the determinant

$$\begin{vmatrix} 1 & z_1 & \cdots & z_1^{p-1} \\ 1 & z_2 & \cdots & z_2^{p-1} \\ \vdots & \vdots & \ddots & \vdots \\ 1 & z_p & \cdots & z_p^{p-1} \end{vmatrix}$$

must be zero. But this is a Vandermonde determinant and implies that

$$z_i - z_j$$

must be zero for some i, j. Or

$$e^{\lambda_i \Delta} = e^{\lambda_j \Delta} ,$$

as required.

REMARK. If A is real, then (9.3) *can* hold for a pair of complex eigenvalues.

$$\lambda_1 = \sigma_1 + i\omega_1 ,$$

$$\lambda_2 = \sigma_1 - i\omega_1 ,$$

provided

$$\omega_1 \Delta = n\pi ,$$

and this is what we have in the example. In general therefore Δ has to be very special before this phenomenon occurs. Note that in sampling band-limited signals, the sampling interval Δ is taken small enough for this not to happen.

We shall not go into the "discrete-time" analogs of all the properties of continuous-time systems, many of them being straightforward. Let us examine, however, the important questions of state estimation and state stabilizability.

THEOREM 9.5. Suppose A is nonsingular and $(C \sim A)$ is observable.

Let $\sigma > 1$ be such that $(\sigma A)^{-1}$ is stable. Let R be the unique solution of

$$R = (\sigma A^*)^{-1} R (\sigma A)^{-1} + C^* C . \tag{9.4}$$

Then R is nonsingular, and defining

$$P = R^{-1} ,$$

we have that

$$(I - PC^* C)A$$

is stable; in fact, all its eigenvalues are less than $1/\sigma$ in magnitude.

PROOF. It is easy to verify that because $(\sigma A)^{-1}$ is stable, (9.4) has the solution

$$R = \sum_{k=0}^{\infty} (\sigma A^*)^{-k} C^* C(\sigma A)^{-k} .$$

Next,

$$Rx = 0 \Rightarrow C(\sigma A)^{-k} x = 0 \qquad \text{for every integer } k \geq 0 .$$

Now $(C \sim A)$ observability implies $(C \sim A^{-1})$ observability (cf. Problem 2). Hence $x = 0$, or R is nonsingular.

Multiplying both sides of (9.4) by P from the left, we have

$$I - PC^* C = \sigma^{-2} R^{-1} A^{*-1} RA^{-1} .$$

Hence

$$(I - PC^* C)A = \sigma^{-2} R^{-1} A^{*-1} R .$$

Let x be an eigenvector of the matrix on the left with eigenvalue γ. Then

$$\sigma^{-2} R^{-1} A^{*-1} Rx = \gamma x$$

or

$$(\sigma A^*)^{-1} Rx = (\sigma\gamma) Rx .$$

Hence

$$\sigma|\gamma| < 1$$

or

$$|\gamma| < \frac{1}{\sigma},$$

as required.

COROLLARY. (Asymptotically optimal state estimate.) Defining

$$\hat{x}_{n+1} = A\hat{x}_n + Bu_n + PC^*(v_{n+1} - C(A\hat{x}_n + Bu_n)) ,$$

we have that

$$\|\hat{x}_n - x_n\| \to 0 \qquad \text{as } n \to \infty .$$

PROOF. Let

$$z_n = \hat{x}_n - x_n .$$

Then from

$$\hat{x}_{n+1} = A\hat{x}_n + Bu_n + PC^*(Cx_{n+1} - C(A\hat{x}_n + Bu_n)) ,$$

$$x_{n+1} = Ax_n + Bu_n ,$$

we have, by subtraction,

$$z_{n+1} = Az_n + PC^*CA(x_n - \hat{x}_n) = (I - PC^*C)Az_n ;$$

and since $(I - PC^*C)A$ is stable, if follows that $\|z_n\| \to 0$ as $n \to \infty$, as required.

Stabilizability

For the state equation

$$x_{n+1} = Ax_n + Bu_n$$

we may obtain a feedback control

$$u_n = -B^*PAx_n$$

so that

$$x_{n+1} = (I - BB^*P)Ax_n ,$$

where

$$(I - BB^*P)A \qquad \text{is stable} .$$

Proceeding analogously to the state-estimator case, we define R by

$$R = (\sigma A)^{-1}R(\sigma A^*)^{-1} + BB^* , \tag{9.5}$$

which, assuming that A is nonsingular and that for some $\sigma > 1$,

$$(\sigma A)^{-1}$$

is stable, has the solution

$$R = \sum_0^\infty (\sigma A)^{-k} BB^* (\sigma A^*)^{-k} .$$

Assuming also that $(A \sim B)$ is controllable, we have that R is nonsingular since $(A^{-1} \sim B)$ is controllable. Define

$$P = R^{-1} .$$

Then, using (9.5),

$$(I - BB^* P) = \sigma^{-2} A^{-1} R A^{*-1} R^{-1}$$

and hence

$$(I - BB^* P)A = \sigma^{-2} A^{-1} R A^{*-1} R^{-1} A .$$

Suppose

$$(I - BB^* P)Ax = \gamma x .$$

Then

$$(\sigma A^*)^{-1} R^{-1} Ax = (\sigma\gamma) R^{-1} Ax .$$

Hence

$$|\sigma\gamma| < 1$$

or

$$|\gamma| < \frac{1}{\sigma},$$

as required.

Time-Varying Systems

We can clearly generalize (9.1) to time-varying systems:

$$\left.\begin{aligned} x_{n+1} &= A_n x_n + B_n u_n , \\ v_n &= C_n x_n + D_n u_n . \end{aligned}\right\} \tag{9.6}$$

Such a representation can arise in periodic sampling of the response of a continuous-time time-varying system; or also if the sampling is nonuniform (nonperiodic) even if the continuous-time system is time-invariant.

Let the sampling epochs be denoted $\{t_n\}$. Let

$$x_n = x(t_n)\,,$$

$$v_n = v(t_n)\,,$$

$$B_n = B(t_n)(t_{n+1} - t_n)\,,$$

$$C_n = C(t_n)\,,$$

where

$$\left.\begin{aligned} \dot{x}(t) &= A(t)x(t) + B(t)u(t)\,, \\ v(t) &= C(t)x(t) + D(t)u(t)\,. \end{aligned}\right\} \tag{9.7}$$

Then letting $\Phi(t)$ be the fundamental matrix solution of

$$\dot{\Phi}(t) = A(t)\Phi(t)\,; \qquad \Phi(0) = I\,,$$

we have

$$x_{n+1} = \Phi(t_{n+1})\Phi(t_n)^{-1}x_n + \int_{t_n}^{t_{n+1}} \Phi(t_{n+1})\Phi(s)^{-1}B(s)u(s)\,ds\,. \tag{9.8}$$

We let

$$A_n = \Phi(t_{n+1})\Phi(t_n)^{-1}$$

and if $(t_{n+1} - t_n)$ is sufficiently small, the second integral in (9.8) can be approximated as

$$B(t_n)u_n(t_{n+1} - t_n)\,,$$

where

$$u_n = \frac{1}{t_{n+1} - t_n}\int_{t_n}^{t_{n+1}} u(s)\,ds \approx u(t_n)\,,$$

and hence we obtain the form (9.6).

We can "solve" (9.6) (analogous to (9.2)) as:

$$x_n = \Psi_n\Psi_m^{-1}x_m + \sum_{m}^{n-1} \Psi_n\Psi_{k+1}^{-1}B_k u_k\,, \tag{9.9}$$

where

$$\psi_n = \Phi(t_n) .$$

Note that if we write

$$\psi(n, m) = \psi_n \psi_m^{-1} ,$$

we obtain the discrete-time analogue of the "transition matrix"

$$\psi(n, m) = \psi(n, k)\psi(k, m) .$$

We can also extend the controllability, observability definitions to (9.6), but characterizing them does not take us too far from the definitions. Thus, as before, we say the state space is *controllable* if it is possible to go from any given state at some epoch t_m to any other state at *some* later time t_n using an appropriate control. Similarly the state space is *observable* if the zero-input response

$$v_n \equiv 0 , \qquad n \geq m ,$$

implies that the state x_m at t_m is zero. Here m is arbitrary.

It is easily deduced that the state space is controllable if and only if the matrix

$$\sum_m^n \psi_k^{-1} B_k B_k^* \psi_k^{*-1}$$

is nonsingular for some $n > m$, for all m. Similarly observability is equivalent to the matrix

$$\sum_m^n \psi_k^* C_k^* C_k \psi_k$$

being nonsingular for some $n > m$, for all m.

Problems for Chapter 9

1. Consider the system

$$x_{n+1} = Ax_n + y_n ,$$

$$y_{n+1} = y_n ,$$

$$v_n = Cx_n .$$

Determine necessary and sufficient conditions for Observability.

2. Suppose A is nonsingular. Show that $(C \sim A^{-1})$ is Observable if $(C \sim A)$ is.

3. Show that (9.1), with $D = 0$, can be expressed in the form

$$x_{n+1} = B_n u_n ,$$

$$v_n = C_n x_n ,$$

and find B_n, C_n.

4. Formulate and solve the "discrete-time" analogue of Problem 3 of Chapter 6.

5. Discretize the continuous-time model in Problem 3 of Chapter 1.

References for Chapter 9

[1] Levan, N. *Systems and Signals*. 2nd ed. New York: Optimization Software, Inc., Publications Division, 1987.

[2] Ogata, K. *Discrete-time Control Systems*. Englewood Cliffs, N.J.: Prentice-Hall Press, 1987.

CHAPTER 10
NONLINEAR SYSTEMS

In this final chapter we present a brief introduction to the study of nonlinear systems in state-space formulation to round out state-space theory. The emphasis is on the concept of asymptotic stability.

Nonlinear Systems: Examples

So far we have only discussed linear systems—in which both the state input equations and the output state equations are linear. Despite much effort we do not as yet possess a state-space theory of nonlinear systems—of systems not necessarily assumed to be linear—matching that for linear systems. Extending the controllability, observability and stability results to nonlinear systems turns out to be much more complex. In particular, there appear new phenomena not present in linear systems.

A general mathematical description of nonlinear dynamic systems may be obtained by extrapolation from (3.1) and allowing all functions therein to be nonlinear. Thus we can write:

STATE INPUT EQUATION

$$\dot{x}(t) = f(t, x(t), u(t)) \,. \tag{10.1}$$

STATE OUTPUT EQUATION

$$\mathrm{v}(t) = C(t, x(t), u(t)) \,. \tag{10.2}$$

Even though this is not exhaustive by any means, the degree of generality is already too much; few, if any, useful results can be deduced, unless additional simplifying assumptions are made. It is instructive therefore to examine first some examples of nonlinear systems arising in practice.

In control systems a common example is where an actuator saturates at high amplitudes. In other words, in

$$\dot{x}(t) = Ax(t) + Bu(t), \tag{10.3}$$

where the control $u(\cdot)$ is feedback so that the linear form

$$u(t) = -B^*P(x(t)) \tag{10.4}$$

must be modified to

$$u(t) = f(x(t)), \tag{10.5}$$

where $f(\cdot)$ has the form

$$\begin{aligned} f(x(t)) &= -B^*Px(t) && \text{for } \|B^*Px(t)\| \le M, \\ &= -M\frac{B^*Px(t)}{\|B^*Px(t)\|} && \text{for } \|B^*Px(t)\| > M. \end{aligned}$$

An iron-core inductor that saturates at high current would be a similar example in electrical engineering.

A canonical example of nonlinearity in mechanical systems is provided by nonlinear damping [5]. Thus, in one dimension we may generalize (7.9) to

$$\frac{d^2x}{dt^2} + 2\zeta\frac{dx}{dt} + f\left(\frac{dx}{dt}\right) + \omega^2 x(t) = u(t), \tag{10.6}$$

where

$$f\left(\frac{dx}{dt}\right) = \gamma\left(\frac{dx}{dt}\right)^3,$$

or, more generally,

$$f\left(\frac{dx}{dt}\right) = \gamma\left(\frac{dx}{dt}\right)^{2n+1}\left(\left|\frac{dx}{dt}\right|\right)^{\beta}, \tag{10.7}$$

where

n is a nonnegative integer, $0 < \beta < 1$,

and γ is a "small" constant ("low" damping).

As an example of a different sort, arising in tracking moving targets, the target dynamics may be described by:

$$\ddot{x}(t) = -\mu\frac{\underline{x}(t)}{\|x(t)\|^3} + u(t)\frac{\underline{x}(t) \otimes \underline{\dot{x}}(t)}{\|x(t)\| \; \|\dot{x}(t)\|},$$

where $x(t)$ denotes the position in a rectangular coordinate system (3×1 matrix) and ⊗ denotes vector cross-product; μ is a constant and $u(\cdot)$ represents the input. The output may be described by

$$v(t) = \begin{vmatrix} \arcsin \frac{x_1}{\|x\|} \\ \\ \arctan \frac{x_2}{x_3} \end{vmatrix} , \tag{10.9}$$

where

$$x(t) = \begin{vmatrix} x_1(t) \\ x_2(t) \\ x_3(t) \end{vmatrix} , \qquad \|x\| = \sqrt{x_1^2 + x_2^2 + x_3^2}$$

(corresponding to measuring "angles only").

Finally we may mention a class of nonlinear systems "closest" (mathematically) to linear systems—the so-called "bilinear" systems, in which the state input equation has the form:

$$\dot{x}(t) = Ax(t) + Bu(t) + \mathcal{B}(u(t), x(t)) , \tag{10.10}$$

where

$$\mathcal{B}(u, x)$$

is linear in u for fixed x and linear in x for fixed u:

$$\mathcal{B}(\alpha u, x) = \alpha\mathcal{B}(u, x) ,$$

$$\mathcal{B}(u, \alpha x) = \alpha\mathcal{B}(u, x) ,$$

$$\mathcal{B}(u+v, x) = \mathcal{B}(u, x) + \mathcal{B}(v, x) ,$$

$$\mathcal{B}(u, x+y) = \mathcal{B}(u, x) + \mathcal{B}(u, y) .$$

We may look upon this as obtained by stopping at the mixed term in a Taylor expansion of the state input equation (10.1).

One of the problems in dealing with nonlinear systems is that one cannot assume that every state input equation one writes down has a unique solution for every initial condition—that, in other words, the state at any time t is determined uniquely by the initial state and the input up to time t. Thus the one-dimensional equation

$$\dot{x}(t) = x(t)^{2/3}$$

has the two solutions:

$$x(t) = \left(\frac{t}{3}\right)^3 , \qquad t \geq 0 ,$$

as well as

$$x(0) = 0 , \qquad t \geq 0 ,$$

both satisfying the same initial condition

$$x(0) = 0 .$$

The reason for this is the fact that the function

$$f(x) = x^{2/3}$$

is not smooth enough. Thus one usually needs to require that the nonlinear functions involved have the adequate degree of smoothness. The strongest such condition is the uniform Lipschitz condition, based on which we may formulate the fundamental theorem on state transition for nonlinear systems described by (10.1).

THEOREM 10.1. Let the function

$$f(t, x, u)$$

satisfy a uniform Lipschitz condition: For any bounded set in t and u, we can find a finite constant M such that

$$\|f(t, x_1, u) - f(t, x_2, u)\| \leq M\|x_1 - x_2\| \qquad (10.11)$$

for all x_1, x_2 in R^n. Then the differential equation

$$\dot{x}(t) = f(t, x(t), u(t)) , \qquad a < t < b ,$$

has a unique continuous solution in $a \leq t \leq b$ for each initial condition $x(a)$ at $t = a$ given and any input $u(\cdot)$ continuous in $a \leq t \leq b$.

PROOF. The basic technique of proof is Picard iteration and may be found in any standard text on differential equations, e.g. [1, 2]. Thus we define

$$x_0(t) = x(a) , \qquad a \leq t \leq b ,$$

and, iteratively,

$$x_n(t) = x(a) + \int_a^t f(s, x_{n-1}(s), u(s))\, ds\ , \qquad a \le t \le b\ .$$

We then show that the sequence $x_n(t)$ converges uniformly on $[a, b]$ to a function $x(t)$ which is a solution. For this purpose let

$$d_n(t) = \|x_n(t) - x_{n-1}(t)\|\ .$$

Then by the uniform Lipschitz condition

$$d_{n+1}(t) \le \int_a^t \|f(s, x_n(s), u(s)) - f(s, x_{n-1}(s), u(s))\|\, ds$$

$$\le M \int_a^t d_n(s)\, ds\ ,$$

since $u(\cdot)$ being continuous on $[a, b]$ is bounded thereon. Now from (10.11)

$$\int_0^t d_n(s)\, ds \le \int_0^t M\left[\int_0^s d_{n-1}(\sigma)\, d\sigma\right] ds$$

$$= M \int_0^t (t - s)\, d_{n-1}(s)\, ds$$

$$\le M^2 \int_0^t (t - s) \left[\int_0^s d_{n-2}(\sigma)\, d\sigma\right] ds$$

$$= M^2 \int_0^t \frac{(t - s)^2}{2}\, d_{n-2}(s)\ ds\ ,$$

and hence by induction

$$\le M^{n-1} \int_0^t \frac{(t - s)^{n-1}}{(n - 1)!}\, d_1(s)\ ds\ .$$

But $d_1(s)$ being continuous on $[a, b]$ is bounded thereon by a constant d; hence

$$d_{n+1}(t) \le M \int_0^t d_n(s)\, ds \le M^n \frac{t^n}{n!}\, d\ .$$

Now

$$\begin{aligned}\|x_{n+p}(t) - x_n(t)\| &\le \|x_{n+p}(t) - x_{n+p-1}(t)\| + \|x_{n+p-1}(t) - x_{n+p-2}(t)\| \\ &\quad + \cdots + \|x_{n+1}(t) - x_n(t)\| \\ &= d_{n+p}(t) + d_{n+p-1}(t) + \cdots + d_{n+1}(t) \\ &\le \sum_{n}^{n+p-1} M^k \frac{t^k}{k!} d ,\end{aligned}$$

and hence it follows that $\{x_n(t)\}$ converges uniformly on $a \le t \le b$ to a continuous function $x(t)$ which satisfies, moreover,

$$x(t) = x(a) + \int_a^t f(s, x(s), u(s)) \; ds$$

and, by differentiation, satisfies the differential equation

$$\dot{x}(t) = f(t, x(t), u(t)) , \qquad a < t < b .$$

Let us next prove that the solution is unique. Let $x_1(t)$, $x_2(t)$ be two solutions. Then

$$x_2(t) - x_1(t) = \int_a^t (f(s, x_2(s), u(s)) - f(s, x_1(s), u(s))) \; ds$$

and letting

$$\|x_2(t) - x_1(t)\| = d_2(t) ,$$

we see that

$$d_2(t) \le M \int_0^t d_2(s) \, ds .$$

This will imply that $d_2(t)$ must be identically zero. For let

$$d_2(t) - M \int_a^t d_2(s) \, ds = g(t) , \qquad a \le t \le b ,$$

and

$$D_2(t) = \int_a^t d_2(s) \, ds .$$

Then we have

$$\dot{D}_2(t) - MD_2(t) = g(t) ,$$

where the super dot indicates derivative with respect to t, and this is a linear differential equation with unique solution in $a \le t \le b$, given by:

$$D_2(t) = D_2(a) + \int_a^t e^{M(t-\sigma)} g(\sigma) \, d\sigma .$$

But $D_2(a)$ is zero and given that

$$g(t) \le 0 , \qquad a \le t \le b .$$

Hence

$$D_2(t) \le 0 , \qquad a \le t \le b ,$$

which is impossible unless

$$d_2(t) \equiv 0 , \qquad a \le t \le b ,$$

or

$$x_2(t) = x_1(t) , \qquad a \le t \le b ,$$

proving uniqueness.

REMARK. The theorem does not apply to the example

$$\frac{dx}{dt} = x^{2/3} ,$$

as we have seen, with $a = 0$. We note also that differentiability in x is not sufficient for uniform Lipschitz property, as the example

$$\dot{x}(t) = e^{x(t)} \tag{10.12}$$

shows. Here

$$e^{x_1} - e^{x_2} = \int_{x_1}^{x_2} e^x \, dx$$

but the derivative e^x is not globally bounded and hence it is **not** true that

$$\|e^{x_2} - e^{x_1}\| \le M \|x_2 - x_1\|$$

for all x_1, x_2. Indeed the differential equation (10.12) has the solution in $0 \le t < 1$ with initial condition $x(0) = 0$:

$$x(t) = \log \frac{1}{1-t} ,$$

which blows up ("explodes") at $t = 1$: a phenomenon without analogue in linear systems. A sufficient condition for uniform Lipschitz property is that $f(t, x, u)$ be continuously differentiable in x and the gradient

$$\nabla_x f(t, x, u)$$

be bounded for bounded t and u. For more details and other aspects of the problem consult [1, 2]. From now on we will simply assume the existence and uniqueness of solution for the state input equation (10.1), and whatever degree of smoothness is needed in proofs.

REMARK. We note that the uniform Lipschitzian condition is satisfied in the bilinear case. But the condition is unnecessarily strong and is often replaced by requiring only that

i) $f(t, x, u)$ is continuously differentiable in x and continuous in all variables;

ii) ("growth condition")

$$\|[x, f(t, x, u)]\| \leq c(1 + \|x\|^2)$$

for bounded t and u.

We note that (i) yields local Lipschitzianness:

$$\|f(t, x_1, u) - f(t, x_2, u)\| \leq M\|x_1 - x_2\|$$

for bounded t, u and x. Under this condition it can be shown that we have existence and uniqueness locally—in some neighborhood of the initial value, and initial time. The growth condition then suffices to show that the solution can be extended to any finite interval $a \leq t \leq b < \infty$. The proof can be found in [1, 2]. Note that the growth condition is not satisfied in (10.12).

Stability

The notions of controllability and observability can be extended in obvious fashion to nonlinear systems, but finding necessary and sufficient conditions characterizing them matching those for linear systems would appear to be impossible. Nor is it possible to obtain a stabilizability or observer theory. More important for us is the notion of stability or "global asymptotic stability" under feedback control. We shall consider the time-invariant case. Thus we specialize to

$$\dot{x}(t) = f(x(t)), \qquad 0 < t, \tag{10.13}$$

$$x(0) \text{ given},$$

where $f(0) = 0$. We shall say that x is an asymptotically stable state if for the solution $x(t)$ corresponding to the initial condition

$$x(0) = x$$

we have that

$$\|x(t)\| \to 0 \qquad \text{as } t \to \infty .$$

We say that the system (10.13) is asymptotically stable, or globally asymptotically stable, if every state is asymptotically stable.

Liapunov Functions

How can we verify asymptotic stability? For a linear system we have the simple criterion that the system matrix A must have all eigenvalues with negative real parts. Moreover, we can find (cf. Theorem 5.3) a nonsingular covariance matrix P such that

$$PA + A^*P + I = 0 .$$

Define now the function

$$V(x) = [Px, x] .$$

This is called a Liapunov function. It has two characteristic properties:

a) $V(x) \to \infty$ if $\|x\| \to \infty$; $V(0) = 0$ if and only if $x = 0$;

b) $[PAx, x] < 0$ for every nonzero x and is zero only if $x = 0$.

We can extend this idea to nonlinear systems. Thus a nonnegative function $V(x)$ is called a Liapunov function for the system (10.13) if

a) $V(x) = 0$ if and only if $x = 0$; $\|V\| \to \infty$ if $\|x\| \to \infty$;

b)
$$[\nabla_x V(x), f(x)] < 0 , \tag{10.14}$$

where $\nabla_x V(x)$ denotes the gradient of $V(x)$ with respect to x:

$$[\nabla_x V(x),\ h] = \frac{d}{d\lambda} V(x + \lambda h)\bigg|_{\lambda=0} .$$

We have then the following theorem.

THEOREM 10.2. Suppose we can find a Liapunov function for the system (10.13) satisfying (a) and (b). Then the system (10.13) is globally asymptotically stable.

PROOF. Let $x(t)$ denote the solution of (10.13) and let

$$\mathrm{v}(t) = V(x(t)) . \tag{10.15}$$

Then

$$\begin{aligned}\dot{\mathrm{v}}(t) &= [\nabla_x V(x(t)), \dot{x}(t)] = [\nabla_x V(x(t)), f(x(t))] \\ &< 0 \qquad \text{if } \|x(t)\| \neq 0 , \\ &= 0 \qquad \text{only if } \|x(t)\| = 0 .\end{aligned}$$

Hence $\mathrm{v}(t)$ is monotone decreasing unless $x(t) = 0$. Let

$$\mathrm{v}(\infty) = \lim_{t \to \infty} \mathrm{v}(t) .$$

We need to show that $\mathrm{v}(\infty)$ is zero. Since $\mathrm{v}(\infty)$ is finite, it follows that $\|x(t)\|$ must be bounded in $0 \leq t < \infty$:

$$\|x(t)\| \leq \mu < \infty . \tag{10.16}$$

Hence we can find a sequence t_k such that $t_k \to \infty$ and $x(t_k)$ converges to x_∞, say. Hence

$$\mathrm{v}(\infty) = V(x_\infty) .$$

We shall now show that x_∞ must be zero. Suppose it is not. Then

$$\|x(t)\| \geq \gamma > 0 \quad \text{for } 0 \leq t < \infty . \tag{10.17}$$

For suppose contrariwise that we can find a sequence $x(t_n)$ such that

$$\|x(t_n)\| \to 0 .$$

If the sequence $\{t_n\}$ is unbounded, we can find a subsequence t_{n_k} such that

$$t_{n_k} \to \infty ; \qquad \|x(t_{n_k})\| \to 0 .$$

But in that case

$$\mathrm{v}(\infty) = V(0) = 0 ,$$

which cannot be, unless $x_\infty = 0$. If the sequence is bounded, we can find a subsequence t_{n_k} such that

$$t_{n_k} \to t_0 \,; \qquad x(t_0) = 0 \,.$$

But this would imply that

$$\mathrm{v}(t_0) = 0$$

and hence that

$$\mathrm{v}(\infty) = 0$$

since

$$\mathrm{v}(\infty) \le \mathrm{v}(t_0) \,.$$

Hence

$$\gamma \le \|x(t)\| \le \mu < \infty \qquad \text{for all } t \ge 0 \,.$$

Now

$$[\nabla_x V(x), f(x)] < 0 \qquad \text{for all } x \ne 0$$

and is a continuous function and hence attains its maximum on the closed bounded set

$$\gamma \le \|x\| \le \mu \,.$$

Let $-\lambda$ denote this maximum. Then

$$\dot{\mathrm{v}}(t) = [\nabla_x V(x(t)), f(x(t))] \le -\lambda \,,$$

or

$$-\dot{\mathrm{v}}(t) \ge \lambda > 0 \qquad \text{for all } t \ge 0 \,.$$

But

$$\mathrm{v}(t) = \mathrm{v}(0) + \int_0^t \dot{\mathrm{v}}(s)\, ds \,,$$

or

$$\mathrm{v}(0) - \mathrm{v}(t) = -\int_0^t \dot{\mathrm{v}}(s)\, ds \ge \int_0^t \lambda\, ds = \lambda t \,,$$

which is a contradiction since $\mathrm{v}(t)$ converges to a finite limit as $t \to \infty$. Hence x_∞ must be zero, proving the theorem.

REMARK. It is not easy to find a Liapunov function. Usually it is a quadratic form:

$$V(x) = [Qx, x], \qquad Q \text{ nonsingular covariance}.$$

Sometimes one may be led to a Liapunov function on physical grounds. For example, in a vibrating system, the total system energy is such a function. Thus let us consider the system:

$$M\ddot{x}(t) + D\dot{x}(t) + Ax(t) + F(x(t), \dot{x}(t)) = 0, \tag{10.17}$$

where A, D, M are nonsingular covariance matrices and $F(x, y)$ is continuously differentiable and is such that

$$F(x, 0) = 0,$$

$$[F(x, y), y] \geq 0. \tag{10.18}$$

The total energy in the system (kinetic plus potential):

$$\mathrm{v}(t) = \frac{1}{2}[M\dot{x}(t), \dot{x}(t)] + \frac{1}{2}[Ax(t), x(t)]. \tag{10.19}$$

Using this function we can show that (10.17) is globally asymptotically stable. Let

$$Y = \begin{vmatrix} y_1 \\ y_2 \end{vmatrix}$$

and define

$$V(Y) = \frac{1}{2}([Ay_1, y_1] + [My_2, y_2]). \tag{10.20}$$

Rewriting (10.17) in the form

$$\dot{Y}(t) = \mathcal{A}Y(t) + \mathcal{F}(Y(t)), \tag{10.21}$$

where

$$\mathcal{A} = \begin{vmatrix} 0 & I \\ -M^{-1}A & -M^{-1}D \end{vmatrix}, \tag{10.22}$$

$$\mathcal{F}(Y) = \begin{vmatrix} 0 \\ -M^{-1}F(y_1, y_2) \end{vmatrix}. \tag{10.23}$$

Note that

$$\nabla_Y V(Y) = \begin{vmatrix} Ay_1 \\ My_2 \end{vmatrix}$$

and hence

$$[\nabla_Y V(Y),\ \mathcal{A}Y + \mathcal{F}(Y)]$$

$$= [Ay_1, y_2] - [y_2, Ay_1] - [Dy_2, y_2] - [F(y_1,y_2), y_2]$$

$$= -[Dy_2, y_2] - [F(y_1,y_2), y_2]\ , \qquad (10.24)$$

which is

$$< 0$$

unless $y_2 = 0$, in which case it is

$$= 0\ .$$

Hence (10.19) would be a Liapunov function for (10.17) except for the fact that (10.24) may be zero without Y being zero—the first component of Y need not be zero for (10.24) to be zero.

In this example however we can exploit the fact that the linear part of the right side of (10.21) is a stable matrix. Hence we can write

$$Y(t) = e^{\mathcal{A}t}Y(0) + \int_0^t e^{\mathcal{A}(t-\sigma)}\ \mathcal{F}(Y(\sigma))\ d\sigma \qquad (10.25)$$

and

$$Y(t) = \begin{vmatrix} x(t) \\ \dot{x}(t) \end{vmatrix} .$$

Since $v(t)$ is monotone decreasing, it follows that $\|x(t)\|$ and $\|\dot{x}(t)\|$ are bounded, and hence so is $\mathcal{F}(Y(t))$, and hence the integral in (10.25) converges. Hence both $x(t)$ and $\dot{x}(t)$ converges as $t \to \infty$. Hence $\mathcal{F}(Y(t))$ converges as $t \to \infty$, and since $\mathcal{A}$ is stable, we can show that the integral in (10.25) must actually go to zero. Let

$$h = \lim_{t\to\infty} \mathcal{F}(Y(t))\ .$$

Then

$$e^{\mathcal{A}t}Y(0) + \int_0^t e^{\mathcal{A}(t-\sigma)}\ h\ d\sigma \qquad (10.26)$$

clearly goes to $-\mathcal{A}^{-1}h$ as $t \to \infty$. The difference between (10.25) and (10.26) is

$$= \int_0^t e^{\mathcal{A}(t-\sigma)}(\mathcal{F}(Y(\sigma)) - h)\, d\sigma ,$$

which we can split up as

$$= \int_0^L + \int_L^t . \qquad (10.27)$$

Given $\varepsilon > 0$, we can choose $L > 0$ such that

$$\|\mathcal{F}(Y(\sigma)) - h\| < \varepsilon \qquad \text{for } \sigma \geq L .$$

Hence the second integral in (10.27) is less in norm than

$$\varepsilon M \int_L^\infty e^{\gamma(t-\sigma)}\, d\sigma , \qquad (10.28)$$

where we use:

$$\|e^{\mathcal{A}t}\| \leq M e^{\gamma t}, \qquad \gamma < 0 ,$$

since $\mathcal{A}$ is stable. Hence (10.28) can be made arbitrarily small, while the first integral in (10.27) goes to zero for fixed L *as* $t \to \infty$. Hence letting

$$Y = \lim_{t\to\infty} Y(t) ,$$

we have:

$$Y + \mathcal{A}^{-1}h = 0 ,$$

or Y must satisfy

$$\mathcal{A}Y = -\mathcal{F}(Y) . \qquad (10.29)$$

But using the fact that

$$F(y_1, y_2) = 0 \qquad \text{for } y_2 = 0 ,$$

we can readily see that $Y = 0$ is the only solution to (10.29). Hence $\|Y(t)\|$ goes to zero, or we have established global asymptotic stability.

Nonlinear Feedback Control

As we have noted, one of the sources of nonlinearity in control systems is the nonlinearity of the controller (actuator). Thus let us consider the system

$$M\ddot{x} + D\dot{x} + Ax + Bu = 0 , \tag{10.30}$$

where for simplicity we shall only consider the case where the control $u(\cdot)$ is one-dimensional. A linear feedback control with desirable robustness properties is (as we saw in Chapter 6):

$$u(t) = B^*\dot{x}(t) .$$

Allowing for saturation of similar effect, we would need to consider in practice

$$u(t) = f(B^*\dot{x}(t)) , \tag{10.31}$$

where

$$f(u) = \text{sign } u . \tag{10.32}$$

Note this function is not "smooth"—it is non-Lipschitz at the origin, and hence solving the differential equation requires special attention. It may however be approximated (for example) by the smooth function:

$$f(u) = \frac{2}{\pi} \tan^{-1} \lambda u , \qquad u \neq 0 , \tag{10.33}$$

for suitably chosen positive constant λ. More generally, we may consider any smooth function $f(\cdot)$ such that

$$uf(u) > 0 . \tag{10.34}$$

The associated stability problem is more generally known as the Lurie problem [3]. The importance of (10.34) is made evident by examining the total energy:

$$E(t) = \frac{1}{2}([M\dot{x}, \dot{x}] + [Ax, x]) .$$

We see that

$$\dot{E}(t) = -[D\dot{x}, \dot{x}] - [Bf(B^*\dot{x}(t)), \dot{x}(t)]$$

and the second term

$$= -[f(B^*\dot{x}(t)),\ B^*\dot{x}(t)]$$

is negative if we impose (10.34). For D nonsingular, we can readily apply Theorem 10.2 to see that the nonlinear feedback system is globally asymptotically stable.

Problems for Chapter 10

1. Show that the one-dimensional system

$$\ddot{x}(t) + 2\zeta\omega\dot{x}(t) + \omega^2 x(t) + \gamma \tan^{-1} \dot{x}(t) = 0,$$

where

$$\zeta, \omega, \gamma \text{ are all } > 0,$$

is globally asymptotically stable. Discuss what happens when $y = 0$ (no damping).

For small enough γ we may approximate the solution by the Krylov-Bogoliubov technique [4]. Thus we write the solution as a damped sinusoid:

$$x(t) = A(t) \sin\left(\omega t + \frac{\pi}{2}\right),$$

where the amplitude $A(t)$ is slowly varying, and essentially constant over each cycle. Let

$$A_k = A(kT),$$

where T is the period:

$$T = \frac{2\pi}{\omega}.$$

Then the approximation yields

$$\log A_{k+1} = \log A_k - 2\pi\zeta - \frac{K_0(A_k)}{\omega A_k},$$

where

$$K_0(a) = \frac{1}{2\pi}\int_0^{2\pi} \gamma \tan^{-1}(a\omega \cos\phi) \cos\phi \, d\phi .$$

Calculate A_k and plot $\log A_k$ as a function of k, the number of cycles. Show that the curve is convex downward (cup). The approximation works for $\zeta = 0$ as well and indeed even replacing

$$\gamma \tan^{-1} \dot{x}(t)$$

by

$$\gamma \operatorname{sign} \dot{x}(t) ,$$

in which case

$$K_0(a) = \frac{1}{2\pi} \int_0^{2\pi} \gamma \operatorname{sign} (a\omega \cos \phi) \cos \phi \, d\phi .$$

2. A typical example of nonlinearity occurs in the dynamics of rotating rigid bodies. Thus, letting ω denote a 3×1 (angular velocity) vector, we have

$$I\dot{\omega} + \omega \otimes I\omega + K\omega + u(t) = 0 ,$$

where I is the moment of inertia matrix, $\otimes$ denotes vector cross-product, K is a nonsingular covariance matrix, and $u(\cdot)$ is the control (moment). Exploiting again the energy function:

$$E(t) = \frac{1}{2} [I\omega(t), \omega(t)]$$

or the Liapunov function

$$V(\omega) = \frac{1}{2} [I\omega, \omega] ,$$

we have

$$\frac{d}{dt} E(t) = [I\dot{\omega}(t), \omega(t)] = -[K\omega(t), \omega(t)] - [u(t), \omega(t)]$$

since

$$[\omega \otimes I\omega, \omega] = 0 .$$

Establish global asymptotic stability for any feedback controller of the form:

$$u(t) = f(\omega(t))$$

such that

$$[f(\omega(t)), \omega(t)] \geq 0 .$$

3. Let x be $n \times 1$ and consider the function:

$$f(x) = \|x\|$$

mapping R^n into R^1. Show that $f(\cdot)$ is Lipschitzian even though $f(\cdot)$ is not differentiable at $x = 0$. Hence show that the differentiable equation in one dimension:

$$\ddot{x} + x + \alpha \dot{x} + \beta \dot{x} |\dot{x}| = 0$$

has a unique stable solution, for $0 < \alpha, \beta$.

References for Chapter 10

[1] Birkhoff, Garrett, and Rota, Gian-Carlo. *Ordinary Differential Equaions*. New York: John Wiley & Sons Publ. Co., 1978.

[2] Hale, J.K. *Ordinary Differential Equations*. New York: Wiley-Interscience, 1969.

[3] Pontryagin, L.S. *Ordinary Differential Equations*. Reading, Mass: Addison Wesley Publ. Co., 1962.

[4] Lee, E.B., and Markus, L. *Foundations of Optimal Control Theory*. New York: John Wiley & Sons Publ. Co., 1968.

[5] Krylov, N.N., and Bogoliubov, N. *Introduction to Nonlinear Mechanics*. Princeton, N.J.: Princeton University Press, 1949.

[6] Nayfeh, A.H., and Mook, D.T. *Nonlinear Oscillations*. New York: John Wiley & Sons Publ. Co, 1979.

INDEX